D#270526

D1123805

WITHDRAWN
JAN 30 2023
DAVID O. McKAY LIBRARY
BYU-IDAHO

# ENERGY AND THE EVOLUTION OF LIFE

# ENERGY AND THE EVOLUTION OF LIFE

▼▼▼

RONALD F. FOX

*School of Physics*
*Georgia Institute of Technology*

W. H. FREEMAN AND COMPANY
NEW YORK

Library of Congress Cataloging-in-Publication Data

Fox, Ronald F. (Ronald Forrest), 1943–
Energy and the evolution of life / Ronald F. Fox.
p. cm.
Includes index.
ISBN 0-7167-1849-9. ISBN 0-7167-1870-7 (pbk.)
1. Evolution. 2. Bioenergetics. I. Title.
QH371.F62 1988 87-21458
575—dc19 CIP



Printed in the United States of America

1 2 3 4 5 6 7 8 9 0 VB 6 5 4 3 2 1 0 8 9 8

*To the memory of*
*Mark Kac and Fritz Lipmann*

# Contents

# Preface

I first began thinking about the role of energy in biological evolution roughly thirty years ago when I was a secondary school student. I would visit my father's laboratory, where I was allowed to make proteinoid microspheres in a test-tube and then watch them under a microscope. I found great fascination in mixing salts, sugars, and lipids and observing the remarkable self-assembled structures that formed. Most captivating of all, however, was their Brownian motion. The microspheres jiggled around, and occasionally a microsphere divided, and I would watch the two offspring move away from each other. However, my aptitudes were strongest in the mathematical sciences, so in college I studied mathematical physics, with the intention of one day returning to biological questions.

During my college years (1960–1964), the genetic code was elucidated, and soon thereafter protein biosynthesis was described in great detail. I was not unaware of these developments because, even though the study of mathematical physics put heavy demands on my time, I kept up with molecular biology. These events greatly influenced my decision to pursue graduate studies at the Rockefeller University (1965–1969), where I studied nonequilibrium thermodynamics with George Uhlenbeck and Mark Kac. I also had the good fortune to learn biochemistry and bioenergetics from Fritz Lipmann and Christian de Duve. At that time, de Duve had the remarkable foresight to realize that the idea of chemiosmosis deserved a serious hearing and he organized a seminar course on chemiosmosis, which I attended. He invited Peter Mitchell and many of his critics (most of whom are now quite converted) to lecture, and I began to realize that a revolution in bioenergetics, no less significant than the revolution in molecular biology, was about to take place. By the mid-1970s this revolution was nearly complete, and today, while many details of bioenergetics require more work (also true in molecular biology), the essential outline is clear.

During the 1970s I pursued an academic career in nonequilibrium statistical mechanics and stochastic processes. I was still looking for chances to connect my studies with biology. While I was a postdoctoral fellow, Uhlenbeck arranged for me to spend a couple of weeks with Max Delbruck to help make this connection. Delbruck was very sympathetic but tried to convince me that what I should do was simply to become a biologist, as he had done so successfully. To him, physics and biology were concerned with very different issues and required very different methods. At about the same time, I had occasion to visit with Lars Onsager, whom I had met as a student at Rockefeller University when he visited Uhlenbeck. Onsager had retired to Miami, where he came to know my father, and revived his earlier interest in questions of the origins of life. I once asked him whether he thought that statistical mechanics would have a significant impact on the study of biological questions. He replied with a typically enigmatic and terse: "No." Ironically, the man who encouraged me the most, when I visited Kac and Uhlenbeck at Rockefeller University, was Fritz Lipmann. After all, he had played the major part in developing our understanding of the role of energy in biology and, while he emphasized the importance of appreciating biochemistry as chemistry, he also understood that energy flow was a physical, rather than a chemical phenomenon with a strongly thermodynamic flavor. Through Lipmann I learned about the unique biological importance of phosphorous compounds and their significance with regard to energy flow.

The 1970s witnessed yet another scientific revolution, this time in mathematical physics, where problems of nonlinear dynamics were being solved and attracting considerable attention. My colleague, Joe Ford, a pioneer in this field during the early 1960s, invited many scientists—among them Boris Chirikov and Mitchell Feigenbaum—to our department to discuss these developments. At first I did not realize what was happening and was mildly amused to hear these people talking about *classical oscillators*, about which I thought I knew everything! As a postdoctoral fellow at Berkeley, I had expressed my dismay at such talk to Karen Uhlenbeck, who patiently attempted to tell me that these oscillators weren't *harmonic* but nonlinear, and that that made all the difference. It took me nearly a decade for these ideas to sink in, so that it wasn't until the beginning of the 1980s that I recognized the point of the new scientific revolution: that deterministic dynamical systems could display chaotic behavior. Even more important for me was the realization that well-posed mathematical systems of equations, which in principle possessed unique solutions, could be successfully studied only in conjunction with numerical simulation. The advent of readily available rapid computing coincided with this revolution.

In 1984, the John Simon Guggenheim Foundation generously awarded me a fellowship to study the physical basis for biological evolution. This award had been supported by Terrell Hill, Mark Kac, Fritz Lipmann, and James Stevenson, to whom I am most grateful. The grant was generously matched by the Georgia Tech Foundation, so that I was able to devote the 1984–1985 academic year exclusively to these studies. My objective was to bring together the consequences of the several revolutions mentioned above. I approached the project as an experiment in simultaneous writing and research. The experience was very rewarding, and this book is the product. I am indebted to the Guggenheim Foundation and to the Georgia Tech Foundation for their support.

I am also especially grateful to a few close friends who were willing to review an early version of the manuscript and provide me with their criticisms. They are Larry Gold, Joel Keizer, and Roger Wartell. I benefitted greatly from the editorial efforts of Barbara Brooks, senior editor for W. H. Freeman and Company, and Martin Silberberg of Manuscripts Associates.

It saddens me to reflect that Mark Kac (1914–1985) and Fritz Lipmann (1900–1986) are no longer able to read my work. I would have loved to have received their comments about this book. Kac served so well as mentor, guide, and friend over the last twenty years. Lipmann encouraged my exploration into the origins of life at several crucial steps. It is to the memory of Kac and Lipmann that I dedicate this book.

*Ronald F. Fox*

# ENERGY AND THE EVOLUTION OF LIFE

# Introduction

Any book about the role of energy in evolution must be an interdisciplinary work, concerned with biology, chemistry, mathematics, and physics. The biological and chemical aspect of the book (Chapters 1, 2, 3, and 5) is descriptive, emphasizing specific substances and properties, whereas the mathematical and physical aspect (Chapters 1, 4, and 5) is conceptual, emphasizing general principles. A glossary of specialized terminology appears at the end of the book.

## Energy Flow in Biological Systems

Biological organization and its evolution are consequences of the flow of energy through matter. Fritz Lipmann's recognition (1941) of the universal energy currency of phosphates in biological systems (see Figure I-1) was essential to the subsequent revolution in molecular biology, a revolution that elucidated the processes of the biosynthesis of proteins and of the genes that serve as templates for them. Scientists mapped metabolic pathways (for example, glycolysis, the tricarboxylic acid cycle, and the electron transport chain) and realized that these pathways are driven by the flow, use, and transduction of energy. This book presents these ideas at an introductory level with emphasis on energy flow as the central principle. The revolution in molecular biology preceded by only a decade a related revolution in bioenergetics, in which scientists discovered that the process of *chemiosmosis* provides the link between electron transport energy and phosphate bond energy (Harold, 1986). This discovery enabled biochemists to describe reasonably well the mechanisms by which cells transduce, use, and store energy.

## Mathematical Properties of Driven Dynamic Systems

The recognition that biological evolution is a manifestation of energy-driven organic matter has focused the attention of biologists on recent revolutionary

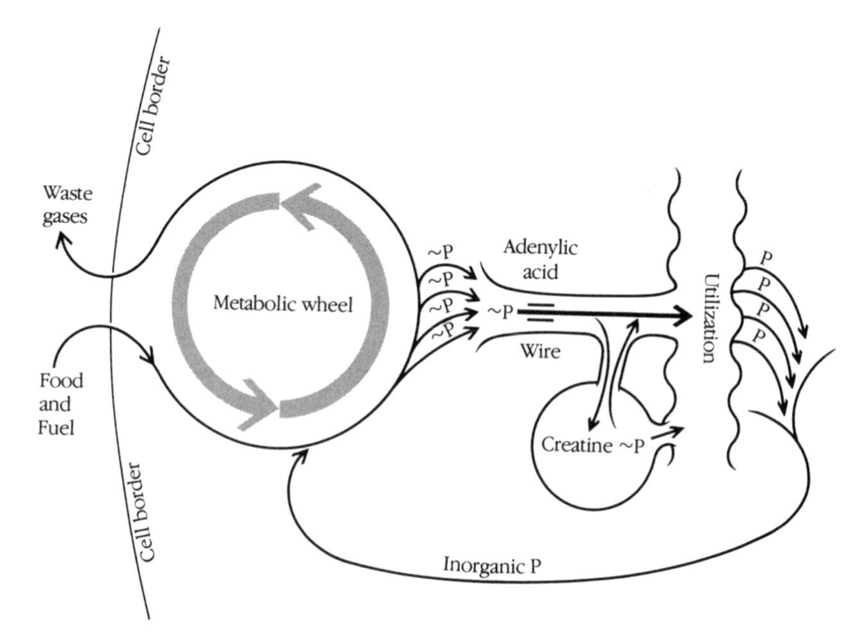


**FIGURE I-1**
The Lipmann cycle. The metabolic dynamo generates ~P current. This is brushed off by adenylic acid, which likewise functions as the wiring system, distributing the current. Creatine~P, when present served as P accumulator. Components of the metabolic wheel include glycolysis, the citric acid cycle, and the electron transport chain.

work by mathematicians and physicists on understanding the dynamics of driven nonlinear systems. The great mathematician, Jules-Henri Poincaré (1854–1912) had gained insights into such systems long before there were computers. However, the systems are best studied by computer simulation because the equations used rarely possess closed-form solutions. By using computers to apply Poincaré's ideas, researchers have discovered many new features of driven nonlinear systems. The concern of this book is with those features—some of which are remarkably simple—that are relevant to the nonlinear biodynamics of evolving organisms.

The discussion of nonlinear dynamics includes an account of the relevance of energy inputs and the remarkably diverse behaviors they produce in driven systems of very simple structures; it also explores the fact that the deterministic equations used to describe driven nonlinear systems can give rise to descriptions of *chaotic* behavior. These behaviors are closely related to the nonexistence of closed-form solutions for these equations and to the need for simulation techniques to obtain information about their dynamics. These considerations have profound significance for the problem of *predictability* in nonlinear systems, which Chapter 5 relates to the process of biological evolution.

## Perspectives on Energy in Biology

This book presents three perspectives on the role of energy in biology, which are related but quite distinct. Chapter 2 gives the first perspective, which concerns the self-generating structures of biological energy-transducing mechanisms and how energy transduction could have happened before any such structure existed. This is the uroboros puzzle. Chapter 3 considers how evolution resulted from sophistications in energy transduction and is concerned with the roles of energy regulation (the second perspective) and storage (the third perspective), especially in certain special energy storage molecules, the phosphagens. Chapters 4 and 5 explore the emergent properties of phosphagen-dependent biodynamics, in particular the excitable tissues, the muscles and nerves, their dependence on phosphagens, and their direct connection with the issues of nonlinear dynamics. The three perspectives of the role of energy in biology cover a range of questions, from the origin of unicellular life to the evolution of multicellular organisms and their social structures. The synthetic overviews that express the conclusions reached from each perspective are the author's hypotheses. Throughout the book, putative facts are clearly differentiated from explicit hypotheses.

***First Perspective: The Uroboros Puzzle*** What is the fundamental problem in biology? Max Delbruck answers as follows: "Thus there is a clear case for the transition on Earth from no-life to life. How this happened is a fundamental, perhaps *the* fundamental question of biology." [Delbruck, 1978, p. 147] The problem of the initiation of life is beautifully embodied in the ancient idea of the uroboros. The uroboros, symbolized by a serpent with its tail in

**FIGURE I-2**
The uroboros. Usually the uroboros is symbolized by one serpent rather than two. This version has a more direct molecular biology interpretation in terms of proteins and polynucleotides. (From Eleazar's *Uraltes Chymisches Werk,* 1760.)

The alchemical uses that grew out of the idea of uroboros were related to Greek ideas about its manifestations in the real world. One of these manifestations was gold, which was equal to fire on the scale of Elements. It seemed to be the only substance that could resist chemical change by fire, and it is yellow like fire and the sun. The sun symbolized gold in ancient literature, and for both the Greeks and the Egyptians, gold color, the spirit of gold was the ideal of perfection. This spiritual essence was endowed with the power to reproduce itself and to transmute all metals into gold. To Plato gold was the ultimate state of inanimate matter. To his pupil, Aristotle, organic nature, also seeking perfection, progresses through a series of cycles in which the ultimate goal is self-reproduction. This view is the strongest analogue to modern molecular biology: The sun, the symbol of self-begetting gold, is the ultimate source of energy for earth's self-begetting systems.

its mouth (see Figure I-2), represents an entity that is self-generating and self-sustaining.

Life itself is a self-generating and self-sustaining system that has evolved into a state of being in which its origins are no longer discernable. Organisms use proteins in energy-transducing structures for obtaining the energy to make the proteins needed for energy transduction. The uroboros puzzle of the transition of no-life to life is: how could this have begun?

***Second and Third Perspectives: Evolutionary Trends*** The uroboros puzzle reflects the fundamental biological problem of the transition from no-life to life. Energy flow supports the existence of life's dynamic molecular state and is the key to evolutionary trends that lead to more complex forms of life, especially multicellular forms that have muscle and nervous tissues. Regulatory mechanisms (perspective two) and storage schemes (perspective three) involve subtle features of energy flow. The solution to the uroboros puzzle pinpoints phosphate bond energy as the special type of energy used by cells. Phosphate compounds are important regulatory substances, and one group, the phosphagens, provides the energy storage that is essential for the function of excitable tissues (muscle and nerves). Phosphate along with calcium, another regulatory substance, provide the substrate for bone, a structural and protective material that has coevolved with excitable tissues.

Max Delbruck (1977) observed that it is difficult to reconstruct the "tree of life from paleontology and comparative anatomy" because fossils provide a spotty record. Although comparative anatomy, physiology, and biochemistry are much more useful than fossils for such a reconstruction, he concludes that "no amount of study of present forms would permit us to infer dinosaurs." However, the evolutionary view of energy flow makes a good case for vertebrates, if not dinosaurs. In particular, muscle, bone, and nervous tissue appear to arise automatically within the perspectives of the evolution of energy use.

## The Evolutionary Significance of Excitable Tissues

Phosphagens enable organisms to accumulate stores of energy for use during periods of high energy need. Early in the history of life, this storage mechanism was a simple regulatory device. But it evolved and made the evolution of excitable tissues possible. Their rapid use of energy could not be sustained without phosphagens.

Muscle provides an organism with movement, not only of the various parts of its body but also of its entire body through its environment. Such movement greatly intensifies organism-environment and organism-organism interactions, which are usually nonlinear in character. The survival of an individual organism or a species depends on the outcome of these interactions.

Nervous tissue provides organisms with the ability to control their movements and to influence the outcomes of biological interactions. The ability to predict outcomes would be an obviously beneficial strategy, but the prediction of nonlinear events is a special type of nonlinear dynamic problem. Chapter 4 shows that the optimal approach to this problem is rapid simulation. That the nervous system has evolved to be a rapid simulator is the subject of Chapter 5.

A list of references follows each chapter. Three references especially useful for their beautiful illustrations and detailed treatments of molecular and cellular biology are Lubert Stryer's *Biochemistry* (1988), Christian de Duve's *A Guided Tour of the Living Cell* (1984), and Franklin M. Harold's *The Vital Force* (1986). The reader will be reminded throughout this book to consult these three titles.

## REFERENCES

Berthelot, M., *Les origines de l'alchimie,* Georges Steinheil, Paris, 1885.

de Duve, C., *A Guided Tour of the Living Cell,* 2 vols., Scientific American Books, New York, 1984.

Delbruck, M., "Mind from Matter?," in XIIIth Nobel Conference, Gustavus Adolphus College, Oct. (1977). *Nature of Life,* Edited by William H. Heidcamp, University Park Press, Baltimore, 1978.

Harold, F. M., *The Vital Force: A Study of Bioenergetics,* W. H. Freeman, New York, 1986.

Hopkins, A. J., *Alchemy: Child of Greek Philosophy,* AMS Press, New York, 1967.

Lipmann, F., "Metabolic Generation and Utilization of Phosphate Bond Energy, in *Advances in Enzymology* 1 (1941): 100–162.

Sivin, N., *Chinese Alchemy: Preliminary Studies,* Harvard University Press, Cambridge, Mass., 1968.

Stryer, L., *Biochemistry,* 3d ed., W. H. Freeman, New York, 1988.

CHAPTER

# 1

# *Energy and Matter: Before Life*

Energy was present in abundance the moment the universe began. Matter came later, forming from energy as the enormous temperatures and pressures of the beginning grew less. This chapter follows the flows of energy and the changes in its form that resulted in conditions appropriate for life. Atoms, molecules, and a suitable surface had to exist before life could begin.

The first steps are clear. After stars form chemical elements in a process called stellar nucleosynthesis, which involves various energy flows (Section 1-

1), the chemical elements combine into small molecules on the relatively low temperature surface of the earth, with help from natural energy sources (Section 1-2). Among these small molecules are the monomers—amino acids, sugars, and bases—which when linked together form the larger molecules essential to life (Sections 1-3 and 1-4).

Not so clear is how to form these larger molecules essential for life on the prebiotic earth, since they tend to break up as fast as they form (Section 1-5). Thus the uroboros puzzle arises: How do you begin to make polymers when polymers themselves are needed to make polymers (Section 1-6)?

## 1-1. ORIGIN OF THE ELEMENTS DURING STELLAR EVOLUTION

The subject of this book begins somewhat after the beginning of the universe, when the stars functioned as factories of chemical elements. The mechanism by which the chemical elements come into existence is *stellar nucleosynthesis.* The processes involved are an example of how *energy flow* produces complex states of matter from simple constituents: A combination of gravitational energy and nuclear energy converts vast quantities of hydrogen gas, the simplest element, into the nuclei of other, more complex, elements. Nucleosynthesis involves nuclear reaction cycles and happens in stages that correlate strongly with changes in stellar structure.

In the *big bang model* of the evolution of the universe, an extremely dense and hot *ylem* expands and cools, giving rise to stars by collapse of local supercritical masses of mixtures of hydrogen and helium (~25 percent helium). During collapse, half the gravitational potential energy of a supercritical mass is converted into kinetic energy, causing the gas temperature to rise; the other half is radiated into space. As the temperature soars, ionization strips electrons from hydrogen atoms, leaving protons.

### The Proton-Proton Reaction

When the temperature of a forming star reaches $10^7$ K and its density is $10^2$ g/cm$^3$ (100 times that of water), the transition to nuclear energy generation begins, with the proton-proton reaction. Below the reaction temperature, the protons (positively charged hydrogen nuclei) repel each other by coulombic forces; but when this repulsion is overcome by a sufficiently energetic collision between two hot protons, then the protons are close enough together for the short-ranged strong nuclear force to act. This nuclear force is strongly attractive and creates a new combination of nuclear particles, with the release of energy in the form of light and/or other particles; that is,

$$p + p \rightarrow (p, n) + e^+ + \nu \qquad \text{(Reaction 1)}$$

in which p denotes a proton, n a neutron, $e^+$ a positron, and $\nu$ a neutrino.

The expression (p, n) denotes the deuteron, a nucleus of mass 2 (the proton has mass 1). In producing a deuteron, the reaction causes one of the protons to undergo a beta (plus) decay into a neutron, positron, and neutrino. (Only a proton bound inside a nucleus can beta decay; a free proton cannot.)

Inside the newly formed star, hot protons surround the deuterons, a condition that leads to another rapid nuclear reaction,

$$(p, n) + p \rightarrow (2p, n) + \gamma \qquad \text{(Reaction 2)}$$

Here $\gamma$ denotes a photon, and (2p, n) denotes the nucleus of helium 3 ($^3$He), which does not combine with protons. Instead, the protons elastically scatter (bounce off each other) without a net change in their combined energy. Deuterons, used in forming $^3$He, are now too rare to combine appreciably with $^3$He. Consequently, the amount of $^3$He increases until hot $^3$He collisions take place:

$$(2p, n) + (2p, n) \rightarrow (2p, 2n) + 2p \qquad \text{(Reaction 3)}$$

in which (2p, 2n) denotes the nucleus of helium 4 ($^4$He, also called $\alpha$ particle and denoted by $\alpha$).

The net overall process, the sum of Reactions (1)–(3), can be written as the multistep nuclear reaction

$$6p \rightarrow \alpha + 2p + 2e^+ + 2\nu + 2\gamma$$

The positrons ($e^+$) eventually annihilate in collisions with electrons ($e^-$) that were previously stripped from hydrogen atoms, thereby producing light ($\gamma$). The neutrinos ($\nu$) tend to flow out and away from the star into interstellar space.

Aside from producing heavier nuclei, these fusion steps release energy. A calculation for the sun will illustrate. The mass of an $\alpha$ particle is only 99.34 percent of the mass of the four protons from which it is produced. Einstein's famous equation $E = mc^2$ computes the amount of energy produced by *nuclear fusion*. It is enormous and dwarfs the contribution from gravitation potential energy. In the sun, 600 million tons of hydrogen fuse every second to form 596 million tons of helium; thus 4 million tons of matter turn into energy. This is equivalent to 800 billion kilowatt-hours (kWh) of energy every second. Nuclear fusion of the sun's available hydrogen will keep it burning at this rate for several billion years; but gravitational energy alone could produce energy at this rate for only 50 million years.

The addition of another nucleon to an $\alpha$ particle leads to unstable products such as $^5$Li and $^5$He, which have lifetimes of $10^{-21}$ s.

As a star ages and its hydrogen changes to helium, its interior becomes nonhomogeneous. In a star's core, where temperature and density are highest, $\alpha$ particles accumulate at the expense of protons. Astrophysical observations indicate that $\alpha$ particles in the core mix only very slowly with a star's outer envelope of protons. The core of α particles grows in size; but at its present temperature ($10^7$ K), the electrostatic repulsion, which is greater between α particles than between protons (because α's are doubly charged), inhibits α-α nuclear fusion reactions. Much higher temperatures are required for the α-α reaction.

Nevertheless, at temperatures of around $10^7$ K, the nuclei of $^3$He and $^4$He can combine to make $^7$Be, which can react in the star to form other isotopes of lithium, boron, and beryllium. Such reactions explain how elements lighter than carbon come into being. They also produce more α's.

The star has now developed a core of hot but nonreactive α's and an outer envelope of protons (ionized hydrogen atoms), electrons, and hydrogen gas. A thin, spherical shell surrounding the α core is the region in which nuclear fusion and energy generation continue. As a consequence, this region is the hottest (around $3 \times 10^7$ K) and the most luminous.

## Origin of Carbon

Gravitational contraction of the core of α's, the next source of energy transformation, again illustrates the creative force of energy flow. As the core contracts, it heats up due to conservation of total energy. The hot core then heats the reactive spherical shell, which expands greatly. Astronomers call stars in this stage of evolution red giants.

When the core α's reach a temperature of $10^8$ K and densities of $10^5$ g/cm$^3$, the coulombic repulsions between α's are overcome, and they can react and fuse. At this stage something peculiar happens. The natural product of α fusion, the beryllium nucleus ($^8$Be), has a lifetime of only $10^{-16}$ s before it reverts to α's. The fusion reaction,

$$2\alpha \rightarrow {}^8\text{Be} + \gamma \qquad \text{(Reaction 4)}$$

does not produce $^8$Be fast enough to dominate the decay of $^8$Be into α's, the reverse of Reaction (4). So how do beryllium and heavier elements form? Salpeter suggested (Fowler, 1967) that perhaps $^8$Be absorbs another α, forming a stable carbon 12 ($^{12}$C) nucleus:

$${}^8\text{Be} + \alpha \rightarrow {}^{12}\text{C} + \gamma \qquad \text{(Reaction 5)}$$

Fred Hoyle suggested (Fowler, 1967) that the short lifetime of $^8$Be meant that Reaction (5) could be fast enough only if a resonant reaction occurred that produced an excited state of the carbon 12 nucleus (which could then decay to the carbon ground state, $^{12}$C):

$${}^8\text{Be} + \alpha \rightarrow {}^{12}\text{C}^* + \gamma \qquad \text{(Reaction 6)}$$

where $^{12}$C* denotes the carbon nucleus in the excited state. The energy and other properties of this state were theoretically computed by Hoyle and subsequently were verified in nuclear experiments.

Stable $^{12}C$ nuclei then react with hot α's to form oxygen nuclei ($^{16}O$), and these react with α's to form neon nuclei ($^{20}Ne$). The process of adding α's to obtain the nuclei of carbon, oxygen, and neon yield less energy per gram of α's than does proton fusion into α's, but it is faster. Consequently, this type of fusion stabilizes the red giant so that no further gravitational collapse occurs at this stage, despite the less efficient conversion of energy.

### Carbon Cycles

Energy from α reactions and from the gravitational collapse of the α core eventually increase the core temperature to $10^9$ K, which is hot enough to drive a different set of nuclear reactions, the *carbon-nitrogen (CN) cycle*; see Reaction (7). This cycle is possible in any star that forms from interstellar gas containing fusion products from supernova explosions of earlier-generation stars. (The CN cycle operates to some extent in the sun at $10^7$ K, but at this lower temperature it is dominated by the more powerful proton-proton reaction.)

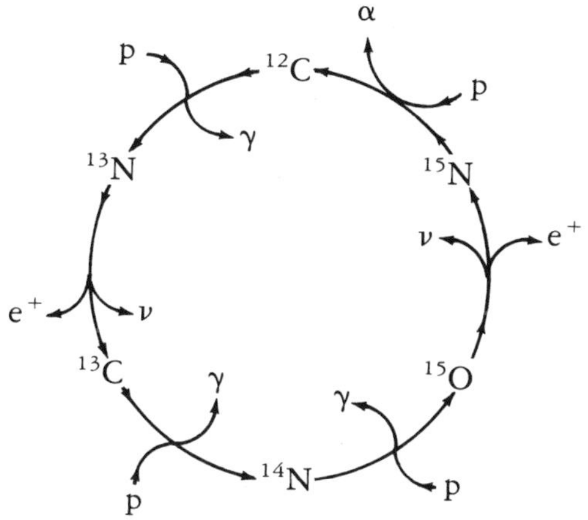

(Reaction 7)

In the CN cycle, carbon acts as a catalyst for the generation of helium from protons. That is, the complex sequence of CN-cycle reactions regenerates $^{12}C$ nuclei as it fuses four protons into one α particle, two $e^+$, two ν, and three γ. The transitions from protons to α's and from α's to $^{12}C$ are both uncomplicated energy-driven conversions. The CN cycle shows for the first time the catalytic effect of an α trimer, $^{12}C$. For approximately every 2000 of these events, an alternative pathway converts $^{15}N$ into $^{14}N$ plus an α through $^{16}O$ and $^{17}O$ intermediates. Both the $^{15}O$ of the CN cycle and the $^{17}O$ are radioactive and decay rapidly; but $^{16}O$ is a stable intermediate that increases in abundance.

If the temperature is high enough ($10^9$ K), $^{12}C$ and $^{16}O$ nuclei react with α's to produce $^{20}Ne$, $^{24}Mg$, $^{28}Si$, and $^{32}S$ (nuclei of neon, magnesium, silicon, and sulfur) as the most stable products. Reactions involving protons (like the reactions in the CN cycle) lead to the intermediate nuclei of fluorine, sodium, aluminum, and phosphorus.

### Production of Heavier Nuclei

Reactions among larger nuclei require higher temperatures because of their larger net charges and their greater coulombic repulsion. However, at $3 \times 10^9$ K the process of photodisintegration begins to contribute to these reactions. The absorption of light by the $^{28}Si$ nucleus causes it to disintegrate into seven α's, which are sometimes captured by another $^{28}Si$ to produce a nickel nucleus ($^{56}Ni$). The $^{56}Ni$ decays into radioactive cobalt ($^{56}Co$), which in turn decays into the stable nucleus of iron ($^{56}Fe$). Many other isotopically related nuclei are also produced in similar processes.

Appearance of the iron group of elements (Co, Ni, and Fe) in some stars suggests that they have achieved nuclear equilibrium at $4 \times 10^9$ K and a density of $3 \times 10^6$ g/cm$^3$. The subsequent evolution of these stars can take one of several different directions that will produce still heavier elements. One outcome is a supernova event, in which the star explodes and distributes the products of its reactions into interstellar space. Later-generation stars will accrete this enriched matter as they form. For the purposes of this book, however, discussion of the evolution of the elements will end with $^{56}Fe$. In the hot star, nuclear reactions occur in a plasma of charged (ionized) nuclei and the electrons that have been stripped from them. At much lower temperatures ($10^4$–$10^5$ K), planet formation occurs concomitantly with atomic element and molecule formation from these electrons and charged nuclei.

### Biological Elements, the Primordial Dozen

The dynamic properties of molecular systems that we recognize as biological are fully represented by chemical reactions among the initial 26 elements (Figure 1-1). Furthermore, for initial inquiries into the transition from no-life to life, the *primordial dozen* (H, C, N, O, Na, Mg, P, S, Cl, K, Ca, and Fe), a subset of the 26 elements, suffices as the basic set from which all biomolecular combinations are constructed. Beyond these, Si and Al are needed for the earth's crust, and several heavier elements—for example, iodine (I), copper (Cu), and zinc (Zn)—are necessary for higher-level nervous tissue functions.

### Summary of Stellar Nucleosynthesis

Gravitational collapse in stars converts gravitational potential energy into heat and light and drives the fusion of protons into α particles. Further collapse and concomitant heating fuses α's into carbon nuclei. These fusion processes

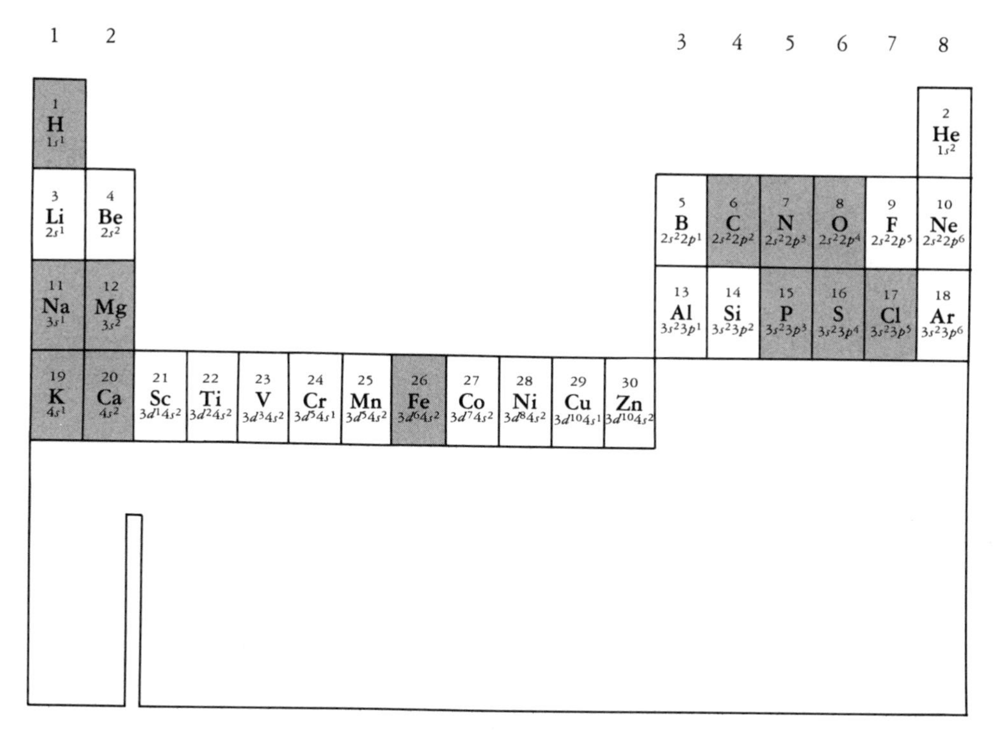


**FIGURE 1-1**
The primordial dozen elements needed for biology are indicated in gray. Some of the others are also important in our treatment, such as aluminum (Al) and silicon (Si) since they contribute to the earth's crust.

release tremendous amounts of energy—much more than from changes in gravitational potential energy. This nuclear energy is created by the conversion of mass into energy. The CN cycle converts protons into $\alpha$'s, with carbon acting as a catalyst. In addition, the $\alpha$'s can react directly with oxygen nuclei to produce the nuclei of neon, magnesium, silicon, and sulfur. In the process of photodisintegration, the energy of light produces even heavier nuclei, including that of iron. Thus, making elements uses four types of energy flow: gravitational potential energy, heat, nuclear energy, and light energy.

## 1-2. FORMATION OF SMALL MOLECULES FROM THE ELEMENTS

The setting of this story now changes from the interiors of stars to the surface of a planet (the earth) that formed around a particular star (the sun) shortly after the solar system was born. On the earth, energy flow at relatively low temperatures will convert the primordial dozen elements into more complex structures: the biologically relevant small molecules.

## Evolution of the Earth

Prevailing models of the evolution of the earth envisage an early stage of accretion about 4.5 billion years ago. Around 4 billion years ago the planet was so hot that the iron-group elements (Co, Ni, and Fe) melted and, because of their great density, passed through the lighter, silicate rock to form the core of the earth. This event is called the *iron catastrophe*. The surface then began to cool and develop a solid crust, which was very thin relative to the earth's radius. The iron-nickel core solidified under great pressure. Current opinion is that the solid iron-nickel core is surrounded by a region of liquid iron and nickel; together these regions account for roughly half the earth's volume. Floating on this liquid iron and nickel is a molten mantle, primarily of silicates, perhaps 2000 km thick. The solid siliceous surface crust is only about 100 km thick, about as thick as an egg shell compared with the diameter of the egg.

Outgassing through the crust built an atmosphere rich in $H_2O$, $N_2$, $CO_2$ and CO. By 3.6 billion years ago, the oceans had begun to form; and great crustal plates, solidified out of the lighter material and floating on the molten interior, were being shaped and moved on the planet's surface by an on-going process called *plate tectonics*. On this early earth with its nascent oceans, temperatures were about $3 \times 10^2$ K and densities were between 1 and 10 g/cm$^3$.

Because light gases such as hydrogen and helium had escaped from the atmosphere, the environment was composed of hydrogen in molecules more complex than $H_2$ and of heavier elements. Table 1-1 shows estimates of the abundances of the elements in various environments including those within organisms. Almost all the hydrogen on the earth and in organisms is bound in water; oxygen is abundant in water, in $CO_2$, and in silicates in the crust. Note for later discussion that the amount of phosphorus in organisms is much greater than in any other environment except the earth's lithosphere.

## Free Energy of Formation, Predictor of Stability

Consideration of the *free energy of formation* ($\Delta G_f^0$) of various molecules gives some idea of what substances to expect on the earth's surface (lithosphere + hydrosphere + atmosphere). By convention, the free energy of formation is zero for the most stable form of every element at 298 K and 1 atmosphere of pressure (that is, $\Delta G_f^0 = 0$). Thus carbon as crystalline graphite has a free energy of formation of 0.0 and is stable; but as diamond it has a positive free energy of formation ($\Delta G_f^0 = 0.685$ kcal/mol) and so is slightly unstable. Table 1-2 presents the empirical formulas, molar weights, and standard free energies of formation for a variety of elements, ions, and biologically important molecules. Because $H_2O$ was abundant on the primitive earth, the properties of many of the ions are given for aqueous solutions (*aq*), which are the biologically relevant forms. (The units kilocalories per mol can be converted to kilojoules per mole by multiplication by 4.186.)

The more negative the free energy of formation the more stable a substance

**TABLE 1-1**
***Abundances of Elements in Different Environments (percent by weight)***

| | | *Earth* | | *Atmosphere* | | |
|---|---|---|---|---|---|---|
| *Element* | *Solar system* | *Lithosphere* | *Hydrosphere* | *Primitive* ($N_2$, $H_2O$, $CO_2$) | *Modern* ($O_2$, $N_2$) | *Organisms (animals)* |
| H | 71.0 | 0.22 | 10.67 | 2.2 | 0.02 | 10.0 |
| He | 27.0 | — | — | — | — | — |
| C | <.01 | 0.19 | 0.01 | 13.3 | 0.01 | 1.67 |
| N | <.01 | 0.02 | 0.02 | 31.1 | 75.53 | 0.18 |
| O | <.01 | 47.33 | 85.79 | 53.3 | 23.02 | 83.94 |
| Na | — | 2.46 | 1.14 | — | — | 0.10 |
| Mg | — | 2.24 | 0.14 | — | — | 0.07 |
| P | — | 0.12 | — | — | — | 1.10 |
| S | — | 0.12 | 0.05 | — | — | 0.14 |
| Cl | — | 0.23 | 2.07 | — | — | 0.16 |
| K | — | 2.46 | 0.04 | — | — | 0.11 |
| Ca | — | 3.47 | 0.05 | — | — | 2.45 |
| Fe | — | 4.5 | — | — | — | 0.01 |
| Si | — | 27.74 | — | — | — | — |
| Al | — | 7.85 | — | — | — | — |

**TABLE 1-2**
***Free Energies of Formation***

| *Name* | *Empirical formula* | *Molar weight (g/mol)* | $-\Delta G^0_{298}$ *(kcal/mol)** |
|---|---|---|---|
| Acetaldehyde | $C_2H_4O$ | 44 | 33.4 |
| Acetic acid | $C_2H_4O_2$ | 60 | 94.7 |
| Acetate (*aq*) | $C_2H_3O_2^-$ | 59 | 89.0 |
| Acetyl CoA | $C_{23}H_{39}O_{18}N_7P_3S$ | 825 | 89.4 |
| Acetyl phosphate | $C_2H_5O_5P$ | 140 | 291.1 |
| *cis*-Aconitate | $C_6H_3O_6^{3-}$ | 171 | 220.0 |
| L-Alanine | $C_3H_7O_2N$ | 89 | 88.7 |
| L-Arginine | $C_6H_{15}O_2N_4$ | 175 | 126.7 |
| L-Asparagine | $C_4H_8O_3N_2$ | 132 | 125.8 |
| L-Aspartate | $C_4H_6O_4N^-$ | 132 | 167.4 |
| Ammonium ion (*aq*) | $NH_4^+$ | 18 | 19.0 |
| Hydrogen carbonate (*aq*) | $CHO_3^-$ | 61 | 140.3 |
| Carbon dioxide (*g*) | $CO_2$ | 44 | 94.2 |
| Carbon dioxide (*aq*) | $CO_2$ | 44 | 92.3 |
| Carbon monoxide (*g*) | $CO$ | 28 | 32.8 |
| Citrate | $C_6H_5O_7^{3-}$ | 189 | 278.7 |
| Creatine | $C_4H_9O_2N_3$ | 131 | 63.1 |
| Creatinine | $C_4H_7ON_3$ | 113 | 6.9 |
| Cysteine | $C_3H_7O_2NS$ | 121 | 81.2 |
| Carbon† (*c*, graphite) | $C_n$ | 12*n* | 0.0 |
| Chlorine† | $Cl_2$ | 71 | 0.0 |
| Chlorine ion (*aq*) | $Cl^-$ | 35.5 | 31.4 |
| Calcium† (*c*) | $Ca_n$ | 40*n* | 0.0 |
| Calcium (*aq*) | $Ca^{2+}$ | 40 | 132.2 |
| Calcium hydrogen phosphate (*c*) | $CaHPO_4$ | 138 | 401.5 |
| Dihydroxyacetone phosphate | $C_3H_7O_6P$ | 170 | 308.9 |
| Erythrose 4-phosphate | $C_4H_9O_7P$ | 200 | 343.8 |
| Ethanol | $C_2H_6O$ | 46 | 43.4 |
| Formaldehyde | $CH_2O$ | 30 | 31.2 |

**TABLE 1-2**
***Free Energies of Formation (continued)***

| *Name* | *Empirical formula* | *Molar weight (g/mol)* | $-\Delta G^0_{298}$ *(kcal/mol)** |
|---|---|---|---|
| Formic acid | $CH_2O_2$ | 46 | 85.1 |
| Formate (*aq*) | $CHO_2^-$ | 45 | 83.8 |
| Fructose | $C_6H_{12}O_6$ | 180 | 218.7 |
| Fructose 6-phosphate | $C_6H_3O_9P$ | 260 | 420.0 |
| Fructose bisphosphate | $C_6H_{14}O_{12}P_2$ | 340 | 621.3 |
| Fumarate | $C_4H_3O_4^-$ | 115 | 144.3 |
| α-D-Galactose | $C_6H_{12}O_6$ | 180 | 220.6 |
| α-D-Glucose | $C_6H_{12}O_6$ | 180 | 219.1 |
| α-D-Glucose 6-phosphate | $C_6H_{13}O_9P$ | 260 | 420.5 |
| L-Glutamate | $C_5H_8O_4N$ | 146 | 166.5 |
| L-Glutamine | $C_5H_{10}O_3N_2$ | 146 | 125.4 |
| Glycerol | $C_3H_8O_3$ | 92 | 116.7 |
| Glycerolphosphate | $C_3H_9O_6P$ | 172 | 319.2 |
| Glycine | $C_2H_5O_2N$ | 75 | 90.0 |
| Glyceraldehyde 3-phosphate | $C_3H_7O_6P$ | 170 | 307.1 |
| Hydroxyl | $HO^-$ | 17 | 37.6 |
| Hydrogen† (*g*) | $H_2$ | 2 | 0.0 |
| Hydronium ion (*aq*) | $H_3O^+$ | 19 | 56.7 |
| Hydrogen peroxide | $H_2O_2$ | 34 | 32.7 |
| Hydrogen sulfide | $H_2S$ | 34 | 6.5 |
| Hydrogen cyanide | HCN | 27 | −28.7 |
| Iron† (*c*) | $Fe_n$ | 55.8*n* | 0.0 |
| Iron(II) (*aq*) | $Fe^{2+}$ | 55.8 | 20.3 |
| Iron(III) (*aq*) | $Fe^{3+}$ | 55.8 | 2.5 |
| Isocitrate | $C_6H_5O_7^{3-}$ | 189 | 277.1 |
| L-Isoleucine | $C_6H_{13}O_2N$ | 131 | 82.2 |
| α-Ketoglutarate | $C_5H_4O_5^{2-}$ | 144 | 190.7 |
| Lactate | $C_3H_5O_3^-$ | 89 | 123.4 |
| α-Lactose | $C_{12}H_{22}O_{11}$ | 332 | 362.0 |

**TABLE 1-2**
***Free Energies of Formation (continued)***

| *Name* | *Empirical formula* | *Molar weight (g/mol)* | $-\Delta G^0_{298}$ *(kcal/mol)** |
|---|---|---|---|
| L-Leucine | $C_6H_{13}O_2N$ | 131 | 85.1 |
| Methane (*g*) | $CH_4$ | 16 | 12.1 |
| Methanol | $CH_4O$ | 32 | 41.9 |
| L-Methionine | $C_5H_{11}O_2NS$ | 149 | 120.2 |
| Magnesium† (*c*) | $Mg_n$ | 24.3*n* | 0.0 |
| Magnesium ion (*aq*) | $Mg^{2+}$ | 24.3 | 109.0 |
| Nitrite ion (*aq*) | $NO_2^-$ | 46 | 8.2 |
| Nitrate ion (*aq*) | $NO_3^-$ | 62 | 26.4 |
| Nitrogen† (*g*) | $N_2$ | 28 | 0.0 |
| Oxalate | $C_2O_4^{2-}$ | 88 | 161.2 |
| Oxaloacetate | $C_4H_2O_5^{2-}$ | 130 | 190.4 |
| Oxygen† (*g*) | $O_2$ | 32 | 0.0 |
| Phosphoric acid | $H_3PO_4$ | 98 | −274.1 |
| Dihydrogen phosphate ion (*aq*) | $H_2PO_4^-$ | 97 | 271.2 |
| Hydrogen phosphate ion (*aq*) | $HPO_4^{2-}$ | 96 | 261.4 |
| Phosphate ion (*aq*) | $PO_4^{3-}$ | 95 | 245.1 |
| Phosphorus† (*c*, white) | $P_n$ | 31*n* | 0.0 |
| Pyruvate | $C_3H_3O_3^-$ | 87 | 113.4 |
| Phosphoenolpyruvate | $C_3H_5O_6P$ | 136 | 303.3 |
| Potassium† (*c*) | $K_n$ | 39.1*n* | 0.0 |
| Potassium ion (*aq*) | $K^+$ | 39.1 | 67.5 |
| Ribose 5-phosphate | $C_5H_{11}O_8P$ | 230 | 382.2 |
| Ribulose 5-phosphate | $C_5H_{11}O_8P$ | 230 | 381.7 |
| Quartz (*c*) (silicon dioxide) | $SiO_2$ | 60 | 192.4 |
| L-Phenylalanine | $C_9H_{11}O_2N$ | 165 | 49.5 |
| Sedoheptulose 7-phosphate | $C_7H_{15}O_{10}P$ | 290 | 457.1 |
| Succinate | $C_4H_4O_4^{2-}$ | 116 | 164.9 |
| Succinyl CoA | $C_{25}H_{40}O_{20}N_7P_3S$ | 882 | 164.0 |
| Sucrose | $C_{12}H_{22}O_{11}$ | 342 | 370.7 |

**TABLE 1-2**
***Free Energies of Formation (continued)***

| *Name* | *Empirical formula* | *Molar weight (g/mol)* | $-\Delta G^0_{298}$ *(kcal/mol)** |
|---|---|---|---|
| Sulfate ion (*aq*) | $SO_4^{2-}$ | 96 | 177.3 |
| Sulfite ion (*aq*) | $SO_3^{2-}$ | 80 | 118.8 |
| Sulfur† (*c*, rhombic) | $S_n$ | $32n$ | 0.0 |
| Silicon (*c*) | $Si_n$ | $28n$ | 0.0 |
| Sodium† (*c*) | $Na_n$ | $23n$ | 0.0 |
| Sodium ion (*aq*) | $Na^+$ | 23 | 62.6 |
| Sodium chloride (*c*) | NaCl | 58.5 | 91.8 |
| L-Serine | $C_3H_7O_3N$ | 105 | 122.1 |
| L-Threonine | $C_4H_9O_3N$ | 119 | 122.9 |
| L-Tyrosine | $C_9H_{11}O_3N$ | 181 | 92.5 |
| L-Tryptophan | $C_{11}H_{12}O_2N_2$ | 204 | 29.9 |
| Urea | $CH_4ON_2$ | 60 | 48.7 |
| L-Valine | $C_9H_{11}O_2N$ | 115 | 86.0 |
| Xanthine | $C_5H_5O_2N_4$ | 153 | 33.3 |
| D-Xylulose | $C_5H_{10}O_5$ | 145 | 178.7 |
| Water | $H_2O$ | 18 | 56.7 |

* 4.186 × kcal/mol = kJ/mol
† Primordial dozen element.

is when it reaches equilibrium. However, the rate at which equilibrium is reached varies over many orders of magnitude for different substances. Consequently, equilibrium considerations are only suggestive, not definitive. Nevertheless, $H_2O$ is a prominent product, given its free energy of formation of −56.7 kcal/mol; and quartz ($SiO_2$), a major crustal component, has a free energy of formation of −192.4 kcal/mol. Calcium phosphate ($CaHPO_4$) is also highly stable at equilibrium, as are the aqueous ions of magnesium ($Mg^{2+}$) and sodium ($Na^+$).

Thermal energy, expressed by $RT$ where $R$ is the gas constant (1.99 cal/mol·K), sets the standard against which to compare these values. At $T$ = 298 K, $RT$ = 0.59 kcal. Statistical mechanics shows that at equilibrium the relative probability for the presence of a molecule with a free energy of formation $\Delta G^0_{298}$ is given by the *Boltzmann formula* $\exp(-\Delta G^0_{298}/RT)$. Almost every entry in Table 1-2, except the elements themselves, is thermally stable at $T$

= 298 K. (25 °C). Note that molecules of biological significance are stable, particularly amino acids, such as alanine and glycine, and simple carbohydrates, such as fructose and glucose. Therefore the basic building blocks of living matter are the relatively stable and therefore abundant small molecules that spontaneously form from the elements, a fact that is consistent with the observation that such molecules are also found in interstellar space, in meteorites, and on the surface of the moon.

## Possible Primitive Atmospheres

Two views of the atmosphere are shown in Table 1-1, one of which is labeled *primitive*. This particular primitive atmosphere contains primarily $CO_2$, $N_2$, and $H_2O$—not the highly reducing conditions postulated in the Oparin-Urey scenario. Instead of a highly reducing primitive atmosphere, many investigators have now proposed a less reducing, or even nonreducing, primitive atmosphere such as that shown in the table. The point here is not to guess which conditions really existed on the primitive earth, but to point out that a number of possible atmospheres might have given rise to conditions favorable to life. The Oparin-Urey scenario for a primitive atmosphere calls for methane ($CH_4$), ammonia ($NH_3$), and water ($H_2O$), as opposed to carbon dioxide ($CO_2$), $N_2$, and $H_2O$, or other choices. When one of a variety of energy fluxes passes through one of these various mixtures of gases, the outcome is a vast variety of molecules with negative free energies of formation, many of which are listed in Table 1-2. The type of energy flux—for example, sparks, ultraviolet (UV) radiation, and heat—and the initial composition of the atmosphere affect the composition of the product mixture. (Details regarding these experiments can be found in the references.) Of course, if the initial mixture does not have a nitrogen-containing component, no nitrogen-containing products can be formed.

Many of the molecules that can be made by adding energy to gas mixtures like the above are biologically relevant because life arose under similar circumstances. The experimental setting arranged by scientists and the natural setting arranged by stellar and geophysical processes automatically produce the molecules that are most likely to form from the primordial dozen elements. Moreover, a variety of settings in the atmosphere, hydrosphere, and *lithosphere* may be relevant simultaneously. Products from one setting may enter another and thereby enrich it. Exactly what happened 4 billion years ago on the earth may remain unknown, but an understanding of the possibilities is within reach.

## Summary of the Formation of Small Molecules

Elements formed during stellar nucleosynthesis combine into varieties of small molecules at the relatively low temperatures on a planetary crust. Energy fluxes, such as lightning, ultraviolet radiation, and volcanic heat, convert the most stable small molecules, such as $CO_2$, $N_2$, and $H_2O$, into combinations such as sugars and amino acids, which become the building blocks for life.

## 1-3. MONOMERS: THE BUILDING BLOCKS OF LIVING SYSTEMS

Of the many small molecules ($MW \leq 300$) that naturally arise from the effects of energy flux on the primordial dozen elements, the *monomers* are of special interest. These include *amino acids*, *monosaccharides* (sugars), and the purine and pyrimidine *bases*. What distinguishes the monomers and makes them special is that they can link together into larger molecules, the *mixed oligomers*, and into still larger ones, the *polymers*.

The chemical process that forms these linkages, found almost everywhere in biology, is called a *dehydration condensation* because formation of the linkage releases a water molecule. The following reaction shows the linkage of two amino acid monomers, in which the $H_2O$ arises from portions of the carboxyl and amino groups of amino acids 1 and 2, respectively. The product (a dipeptide) contains the important *peptide linkage* of proteins:

$$-\overset{\displaystyle O}{\overset{\|}{C}}-\underset{\displaystyle H}{\underset{|}{N}}-$$

$$H_3N^+-C(R_1)(H)-C(=O)O^- + H(H_2N^+)-C(R_2)(H)-C(=O)O^-$$

Amino acid 1 Amino acid 2

Peptide linkage

$$\rightarrow H_3N^+-C(R_1)(H)-C(=O)-N(H)-C(R_2)(H)-C(=O)O^- \quad + H_2O$$

Dipeptide Water molecule

Another example of dehydration condensation is the linkage of two phosphate molecules to produce *pyrophosphate*:*

* For simplicity, here and in later sections, strict adherence to the correct charge states of molecules at pH 7 is not followed unless it is a crucial consideration.

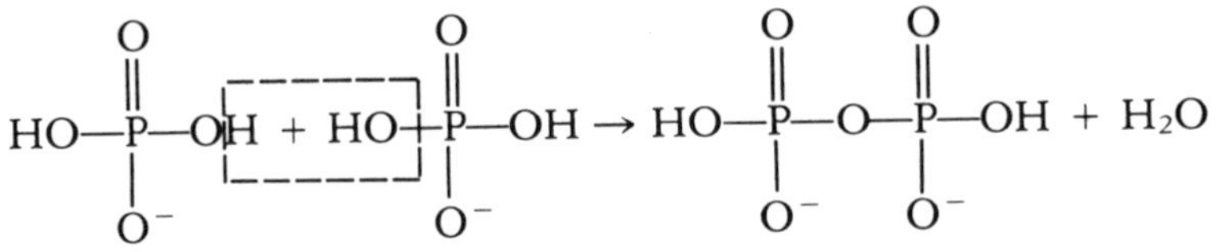

$$\mathrm{HO{-}\overset{\overset{O}{\|}}{\underset{\underset{O^-}{|}}{P}}{-}OH + HO{-}\overset{\overset{O}{\|}}{\underset{\underset{O^-}{|}}{P}}{-}OH \rightarrow HO{-}\overset{\overset{O}{\|}}{\underset{\underset{O^-}{|}}{P}}{-}O{-}\overset{\overset{O}{\|}}{\underset{\underset{O^-}{|}}{P}}{-}OH + H_2O}$$

Phosphate Phosphate Pyrophosphate Water

## Amino Acids, the Monomers of Proteins

When researchers mix $CH_4$, $H_2O$, and $NH_3$ and make them react by applying an energy flux of electric discharge (Miller, 1953) or of heat (Harada and Fox, 1964), the elements H, C, N, and O combine. Amino acids are among the resulting molecular species. Their general structure, in aqueous milieu, is

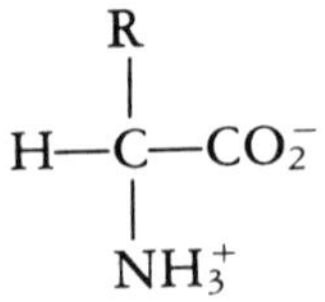

$$\mathrm{H{-}\overset{\overset{R}{|}}{\underset{\underset{NH_3^+}{|}}{C}}{-}CO_2^-}$$

Note that (at room temperature $T$ and neutral pH) the carboxyl group, $-CO_2^-$, is negatively charged, and the amino group, $-NH_3^+$, is positively charged. The R refers to R group, the moiety that confers individuality on the specific amino acids, which are otherwise almost identical. Contemporary organisms have more than 100 different R groups, of which they use only 20 to make their proteins. The structures of these special 20 are in any standard biochemistry text [Stryer, 1988]. The dehydration linkage between the amino acids that make up protein chains occurs between the carboxyl and amino groups and does not involve the R groups.

The simplest amino acid, glycine, for which R is only one atom of hydrogen, has a free energy of formation of $-90$ kcal/mol in $H_2O$, whereas its stable constituents ($CO_2$, $NH_3$, and $CH_4$) together have a free energy of formation of $-123$ kcal/mol. Consequently, although glycine is energy-rich relative to its constituent small molecules, it has a negative free energy of formation relative to its constituent elements. Thus glycine is of intermediate thermodynamic stability, as are the other amino acids. Amino acids existed on the primitive earth because the geochemical conditions on its surface were not equilibrium conditions but involved a variety of energy fluxes, which generated molecules of intermediate stability.

## Sugars, the Monomers of Polysaccharides

Melvin Calvin (1950) reacted $H_2$ and $CO_2$ together using an energy flux of radioactivity to produce monosaccharides (sugar monomers), all of which have the general structure $(CHOH)_n$, where $n = 3, 4, \ldots, 7$. For example, glyceraldehyde, the triose (three carbon atoms), has the following structure:

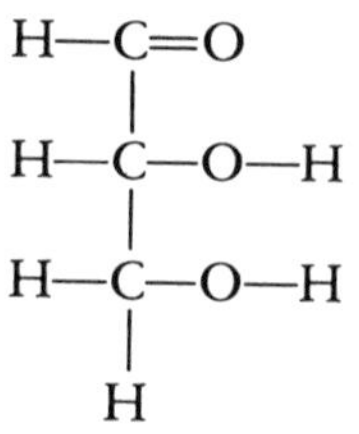

H—C=O
|
H—C—O—H
|
H—C—O—H
|
H

The atoms of $(CHOH)_n$ have many arrangements (called isomers). Some monosaccharides exist largely as closed rings in solution, such as β-D-glucose:

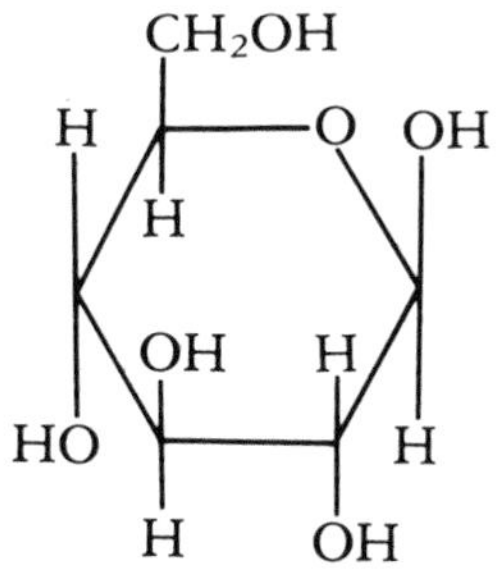

$CH_2OH$
H O OH
H
OH H
HO H
H OH

The hydroxl groups (—OH) in monosaccharides take part in the dehydration linkages that bond the monomers into polysaccharides. Table 1-3 lists the monosaccharides that are important for this discussion. Their structures are in any standard biochemistry text [Stryer, 1988].

The carbohydrate structure $(CHOH)_n$—or, equivalently, $(CH_2O)_n$—has a free energy of formation per $CH_2O$ unit that depends on $n$. Formaldehyde, with $n = 1$, has a free energy of formation of $-31$ kcal/mol, compared with $-149.9$ kcal/mol for $CO_2$ and $H_2O$ (think of $CO_2 + H_2O \rightarrow CH_2O + O_2$). Glucose, with $n = 6$, has $-36.5$ kcal/mol per $CH_2O$ unit. In all cases, the

**TABLE 1-3**
***Simple Sugars***

| *Monosaccharide* | *n* | *Linear* | *Ring* |
|---|---|---|---|
| Glyceraldehyde | 3 | X | |
| Erythrose | 4 | X | |
| Ribose | 5 | X | X |
| Ribulose | 5 | X | |
| Xylulose | 5 | X | |
| Glucose | 6 | X | X |
| Fructose | 6 | X | X |
| Sedoheptulose | 7 | X | |

Adenine (a purine)

Guanine (a purine)

Uracil (a pyrimidine)

6,7-Dimethylisoalloxazine
(part of riboflavin, vitamin $B_2$)

Thiamine

Cytosine (a pyrimidine)

Thymine (a pyrimidine)

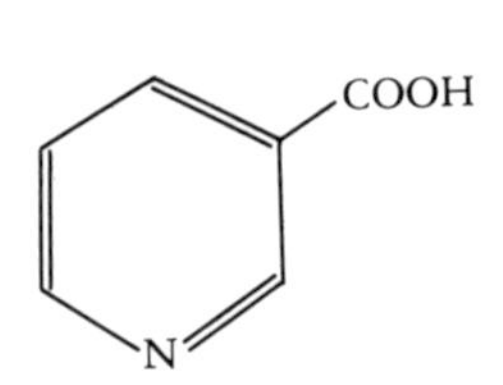


Nicotinic acid (niacin)

**FIGURE 1-2**
Monomers: purines, pyrimidines, and coenzymes.

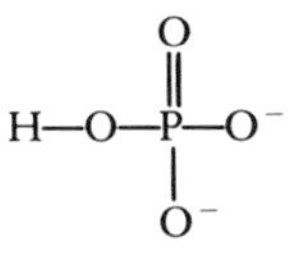

Hydrogen phosphate ion

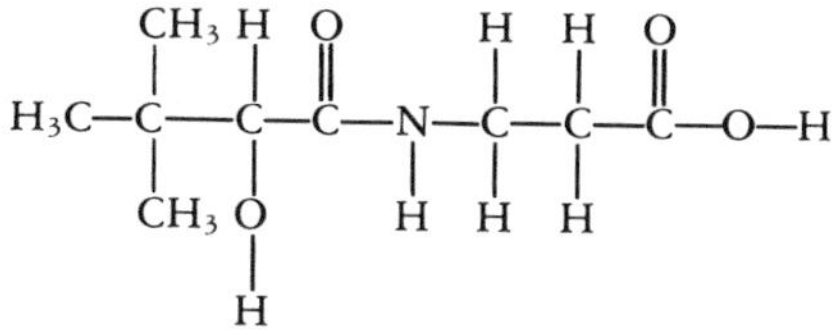

Pantothenic acid

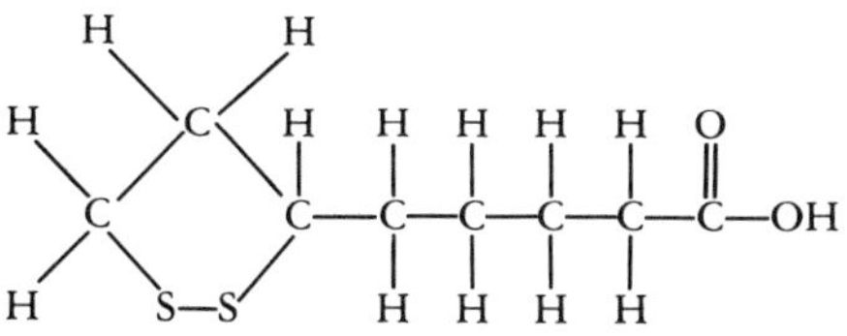

Lipoic acid (oxidized), LSS

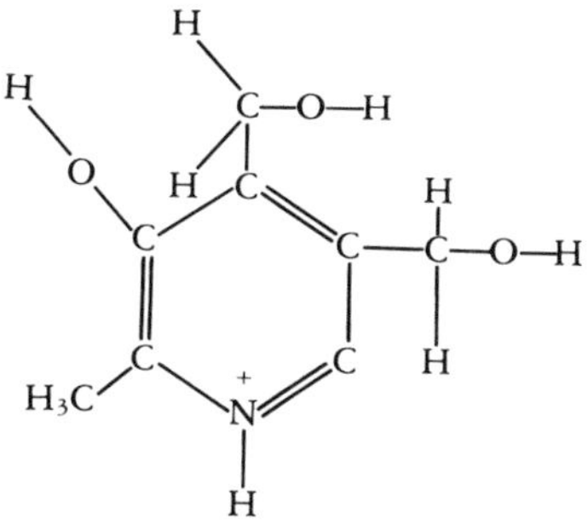

Pyridoxine (part of vitamin $B_6$), PyrP

**FIGURE 1-2** (continued)

saccharide is energy-rich compared with the molecules $CO_2$ and $H_2O$ but is energy-poor compared with the elements themselves. This fact shows that monosaccharides, like the amino acids, are of intermediate stability from an equilibrium point of view. Their existence is a consequence of the presence of energy fluxes.

## Other Small Molecules, the Monomers of Mixed Oligomers

Other molecular species are also regarded as monomers because they can link together by dehydration to produce mixed oligomers. These include many fascinating substances that have special importance. Figure 1-2 displays a representative list. The purines and pyrimidines, for example, are important components of deoxyribonucleic acid (DNA) and ribonucleic acid (RNA) and other species are *coenzymes*, the catalytic centers of many enzymes.

All of these oligomers are of contemporary biological significance, and all are possible products of *abiological* geophysical processes (Oro and Kimball, 1961). For example, $CH_4$, $NH_3$, and $H_2O$ can react in an energy flux of $\beta$ rays to produce adenine or in a flux of electric discharges to produce porphyrin; thus each is a possible constituent of primitive planetary settings. However, their mere presence does not imply that biology is present; it is only a precondition. Biology can become a feature of our picture only after these monomers link into larger structures. The next section explores this possibility and its requirements for energy.

### Summary of the Origin of Monomers

Using a variety of energy fluxes, simple precursors combine to become monomers. In the laboratory in simulated, primitive earth environments that include $CH_4$, $NH_3$, $CH_2O$, and $H_2O$, scientists have produced amino acids, sugars, and the genetic bases.

## 1-4. MIXED OLIGOMERS: SHORT CHAINS OF VARIOUS MONOMERS

This section discusses mixed oligomers (so called because they consist of two to ten monomers that are not all the same kind), which are constituents of enzymes, genes, vitamins, and such important molecules as lactose and sucrose. Subsections treat oligosaccharides, nucleosides, nucleotides, and coenzymes.

Figure 1-3 shows several biologically prominent mixed oligomers and their constituent moieties, each of which combines by the dehydration linkage. Many of these species are coenzymes that are indispensible to energy metabolism in contemporary organisms. Notice that nicotinamide is a dehydration linkage between nicotinic acid and ammonia.

Not all mixed oligomers form in the abiological context of the primitive earth. Section 1-5 addresses their actual occurrence, along with that of the polymers.

### Oligosaccharides

*Oligosaccharides* are comprised of two, three, four, five, six, or more sugar monomers. Sometimes they form as *hydrolysis* products of molecules such as maltose (which has two sugar monomers) or cellobiose (which has two). (In hydrolysis, water inserts itself into a molecule, the reverse of dehydration condensation.) The sugars lactose and sucrose also are oligosaccharides; and sucrose forms as the dehydration condensate of D-glucose and D-fructose:

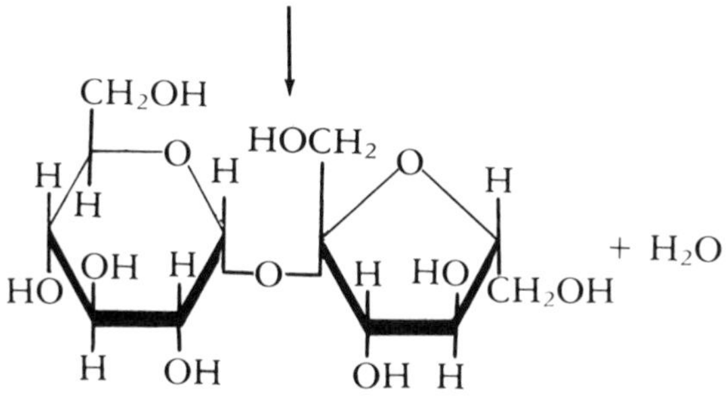

This linkage between the C-1 carbon of glucose and the C-2 carbon of fructose is a special kind of dehydration condensation called a *glycosidic linkage.*

### Nucleosides

*Nucleosides* are oligomers made of a furanose monomer and a nucleic acid base (also a monomer). One molecule of the furanose *ribose* and a molecule of the base *adenine* combine by dehydration condensation to form adenosine:

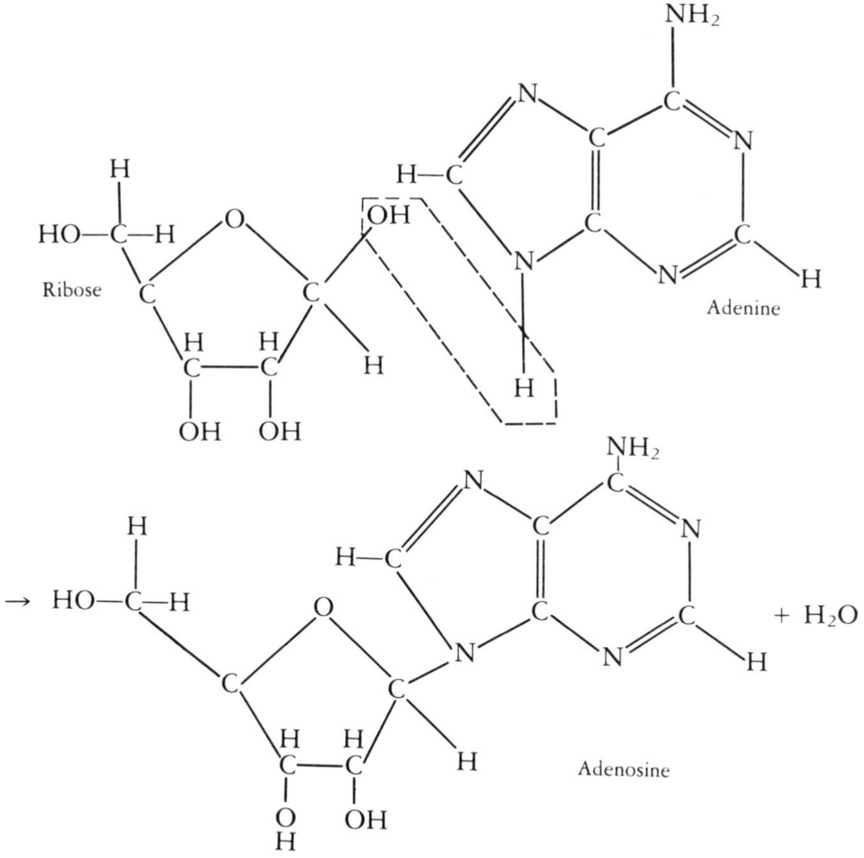

Ribose

Adenine

Adenosine

The addition of a *phosphate* monomer to a nucleoside produces a nucleotide.

1 Nicotinamide mononucleotide (NMN)

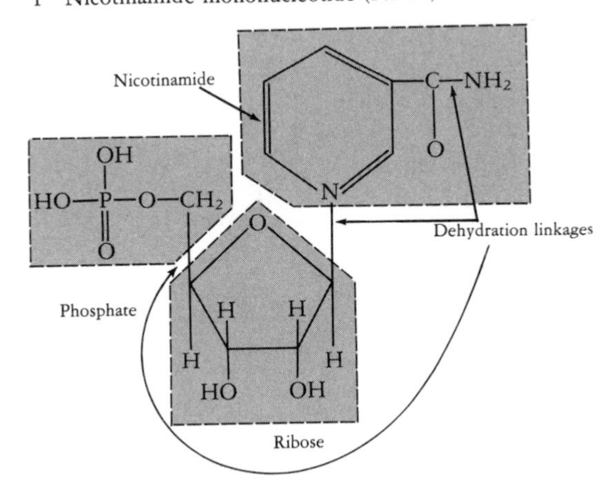


2 Flavin mononucleotide (FMN)

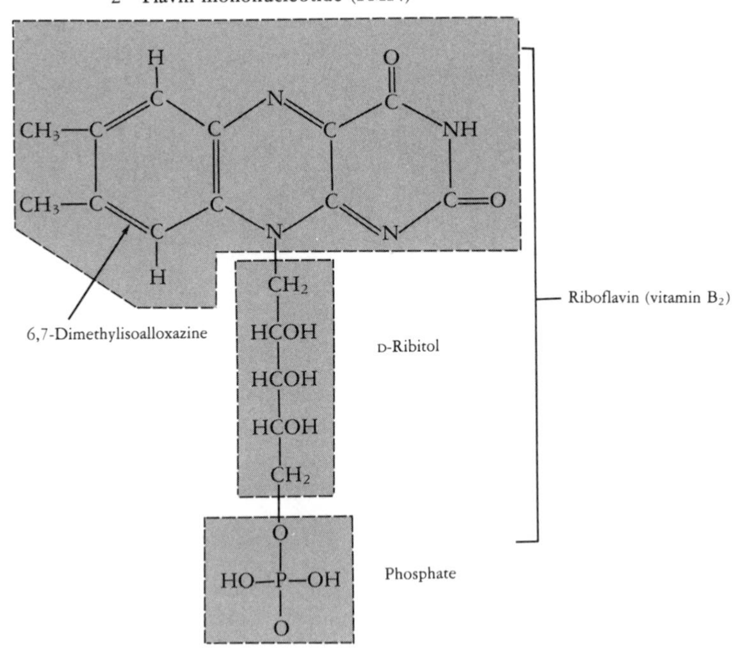


**FIGURE 1-3**
Mixed oligomers.

3 Thiamine pyrophosphate (TPP)

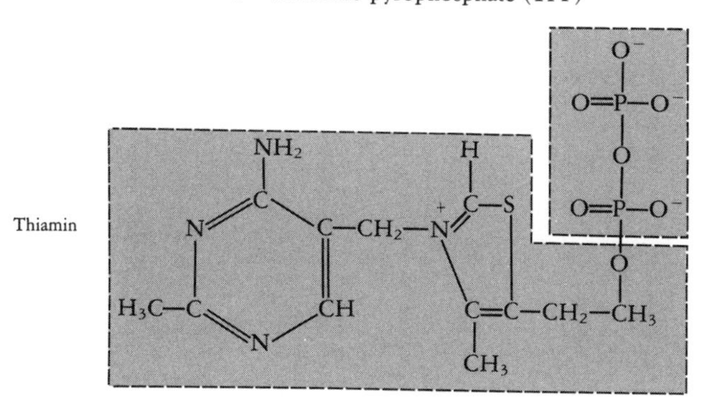


4 Nicotinamide-adenine dinucleotide (NAD)

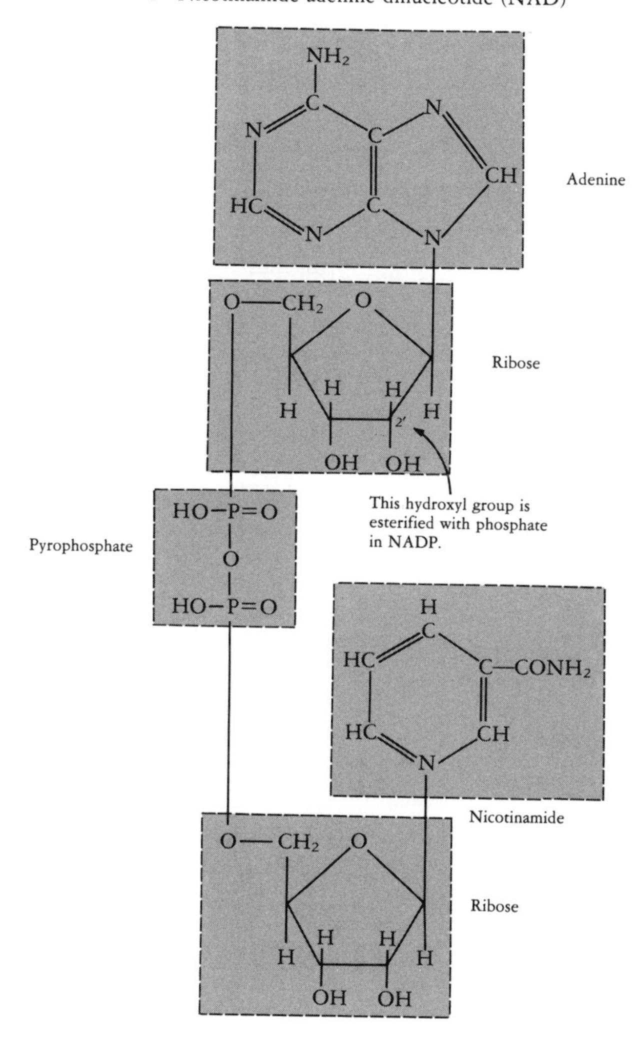


FIGURE 1-3 (continued)

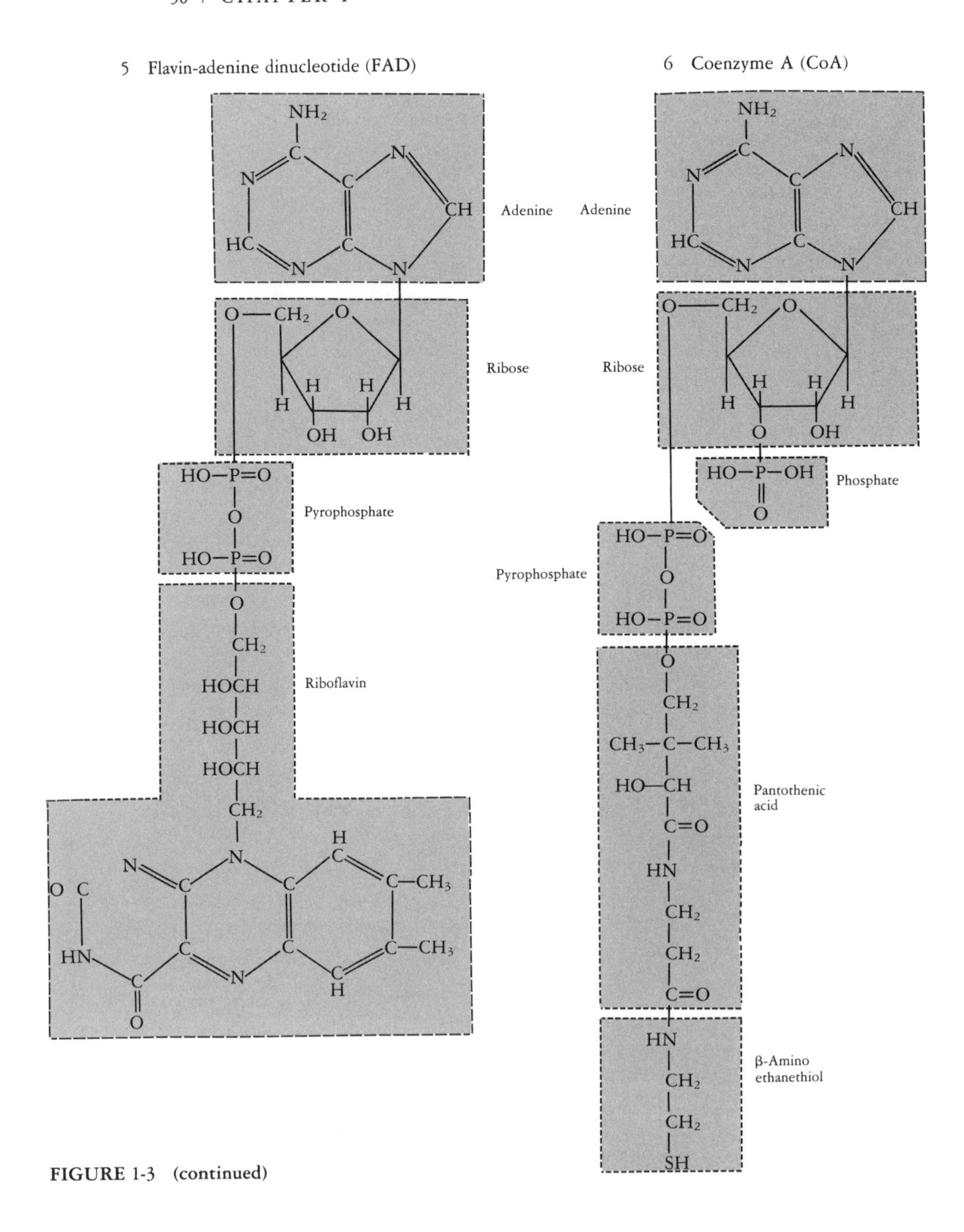


**FIGURE 1-3** (continued)

## Nucleotides

The *nucleotides*, constituents of the genetic system of all organism, are mixed oligomers having three monomers: a nucleic acid base (purine or pyrimidine), furanose (a sugar), and a phosphate (Figure 1-4). The distinction between *ribo-* and *deoxyribo-* in ribonucleic acid (RNA) and deoxyribonucleic acid (DNA) lies in the two types of furanose molecules involved. The three monomers of each nucleotide also bond by dehydration condensation.

Describing nucleotides as mixed oligomers emphasizes their complex structures compared with those of amino acids or monosaccharides; however, nucleotides themselves act as monomers of DNA and RNA macromolecules. Experiments such as those that produce amino acids and monosaccharides in simulated primitive earth atmospheres yield small quantities of nucleic acid bases, but these do not combine into nucleotides, probably because nucleotides require too many dehydration condensations. As the number of dehydration condensations increases, more energy is needed because each linkage requires energy.

## Coenzymes

Coenzymes, also made by dehydration linkage, are of critical importance in metabolic pathways. In contemporary organisms, they combine with proteins (apoenzymes) to form functioning enzymes, which act as catalysts for many biochemical reactions. The coenzyme is the locus of activity of an enzyme, and each coenzyme has a definite biochemical function (see Figure 1-3 and Table 1-4). They participate in a variety of oxidation-reduction reactions and act as carriers of carbon groups. A coenzyme can function in metabolic re-

**TABLE 1-4**
***Coenzyme Functions***

| *Molecule* | *Function* |
|---|---|
| NAD | Hydrogen and electron transfer in oxidation-reduction reactions |
| FAD, FMN | Hydrogen and electron transfer in oxidation-reduction reactions |
| TPP | Decarboxylation of pyruvate and acetyl transfer |
| CoA | "Active" acetyl transfer, "active succinate" |
| PyrP | Codecarboxylase |
| LSS (lipoic acid) | Active acetate transfer, active succinate transfer |
| THF (tetrahydrofolic) | $C_1$-transfer: —CHO (formyl), —HCOOH (formate), —$CH_2OH$ (hydroxymethyl) |
| Glutamic acid | Amine carrier |

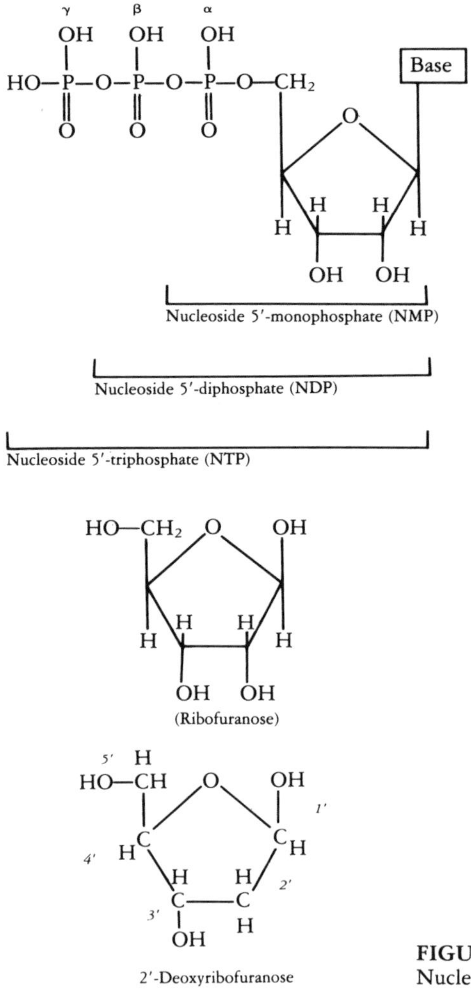

Nucleoside 5′-monophosphate (NMP)

Nucleoside 5′-diphosphate (NDP)

Nucleoside 5′-triphosphate (NTP)

(Ribofuranose)

2′-Deoxyribofuranose

**FIGURE 1-4**
Nucleoside phosphates.

actions even without being joined with a protein to make a complete enzyme; however, the protein is the part that enhances rate of reaction and specificity.

Although not classified as a coenzyme, glutamic acid (an amino acid) acts as a coenzyme while serving as an amine carrier by transferring —$NH_2$ groups. When it gives up —$NH_2$, it becomes α-ketoglutaric acid, an intermediate of the tricarboxylic acid cycle (to be discussed later).

## Summary of Production of Mixed Oligomers

To make mixed oligomers, which are biologically important molecules, monomers link by dehydration condensations. This mechanism for making larger molecules is universal in biology. Energy is required for each linkage. The next section considers how to form polymers (macromolecules).

## 1-5. POLYMERS: THE FABRIC OF LIVING SYSTEMS

All biologically important polymers can be looked on as repeating dehydration condensations between their monomers. However, the mechanisms by which contemporary organisms manufacture polymers are complex multistep processes; simple dehydration does not actually occur. In other words, in the cell, nucleosides are not made by dehydrating ribose and bases, and *polynucleotides* are not made by simply dehydrating nucleotides. Nevertheless, viewing polymers as long chains of monomers linked by dehydrations is convenient.

### Polypeptides and Polysaccharides

Proteins, or polypeptides, are chains of many amino acids that are successively linked by dehydrated peptide bonds:

$$\mathrm{H_2N^+}\!-\!\underset{\mathrm{H}}{\overset{\mathrm{R_1}}{\mathrm{C}}}\!-\!\overset{\mathrm{O}}{\overset{\|}{\mathrm{C}}}\!-\!\underset{\mathrm{H}}{\mathrm{N}}\!-\!\underset{\mathrm{H}}{\overset{\mathrm{R_2}}{\mathrm{C}}}\!-\!\overset{\mathrm{O}}{\overset{\|}{\mathrm{C}}}\!-\!\underset{\mathrm{H}}{\mathrm{N}}\!-\!\underset{\mathrm{H}}{\overset{\mathrm{R_3}}{\mathrm{C}}}\!-\!\overset{\mathrm{O}}{\overset{\|}{\mathrm{C}}}\!-\!\underset{\mathrm{H}}{\mathrm{N}}\!-$$

$$-\underset{\mathrm{H}}{\overset{\mathrm{R_4}}{\mathrm{C}}}\!-\!\overset{\mathrm{O}}{\overset{\|}{\mathrm{C}}}\!-\!\underset{\mathrm{H}}{\mathrm{N}}\!-\!\underset{\mathrm{H}}{\overset{\mathrm{R_5}}{\mathrm{C}}}\!-\!\overset{\mathrm{O}}{\overset{\|}{\mathrm{C}}}\!-\!\underset{\mathrm{H}}{\mathrm{N}}\!-\!\underset{\mathrm{H}}{\overset{\mathrm{R_6}}{\mathrm{C}}}\!-\!\overset{\mathrm{O}}{\overset{\|}{\mathrm{C}}}\!-\!\mathrm{O^-}$$

When joined, the original $n$ amino acids yield this polypeptide and $n - 1$ molecules of water. The reverse process, the degradation of proteins into their constituent amino acids by the action of water, is called hydrolysis. Polysaccharides are long, often branched, chains of monosaccharides, successively linked by glycosidic dehydrations.

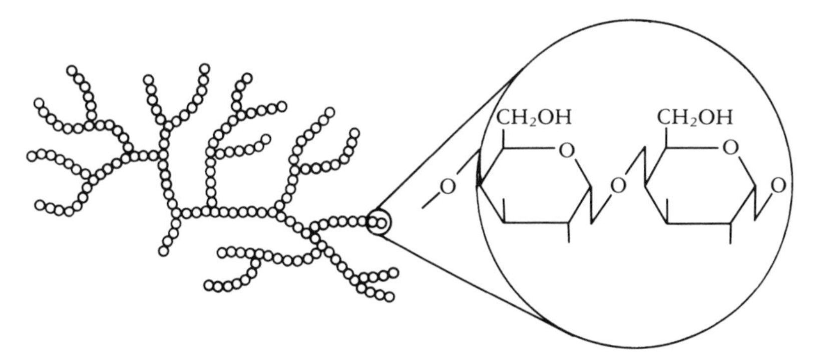


By virtue of their size and the nature of the crosslinks between their chains, some polymers possess important structural properties. For example, callogen is a strong fibrous protein in connective tissue, and chitin, a polymer of the monosaccharide N-acetyl-D-glucosamine is the horny substance in the exoskeletons of crutaceans and insects.

Many proteins, especially the enzymes, are globular in physiological salt solutions. These globular conformations can themselves be viewed as monomers on a higher level of structural hierarchy in which aggregates of protein globules act as oligomers. The aggregation process does not involve dehydration condensations even though water plays a central role. (Chapter 3 contains further discussion of this fascinating property of proteins.)

### Polynucleotides

Polynucleotides (DNA and RNA) are chains of mononucleotides linked by *phosphodiester* dehydrations:

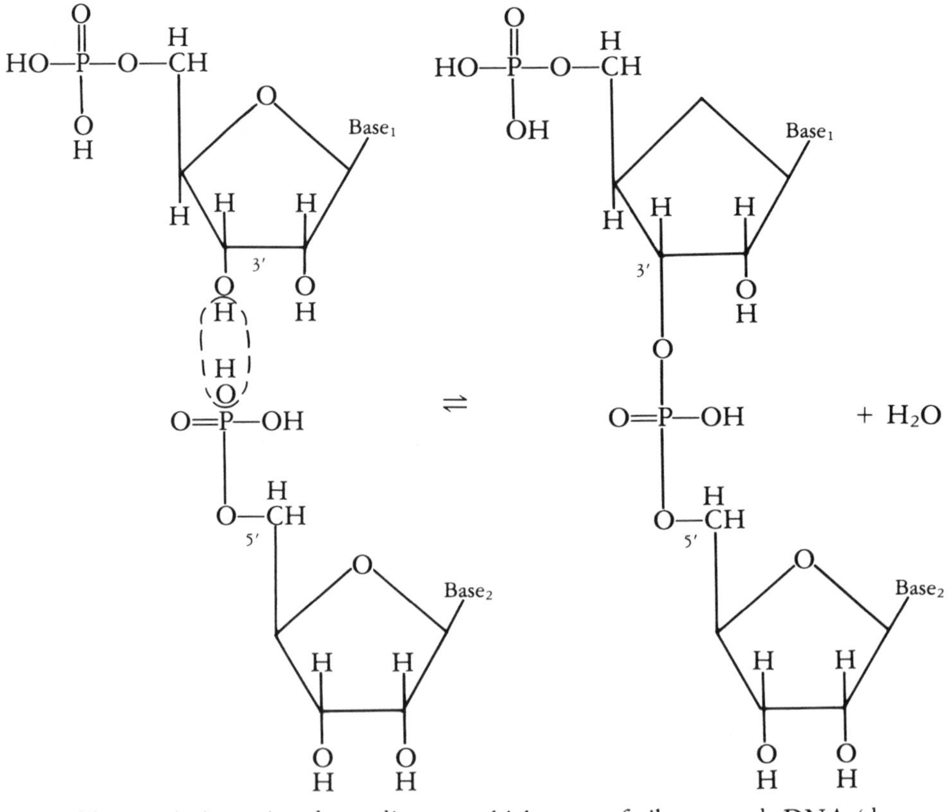

Two varieties exist, depending on which type of ribose used: DNA (deoxyribonucleic acid) and RNA (ribonucleic acid). Also, several functionally different types of RNA exist.

### Energy Requirement for Polymerization

So far the emphasis in this chapter has been on structure. In every case, what organisms must do to achieve these structures involves energy. The en-

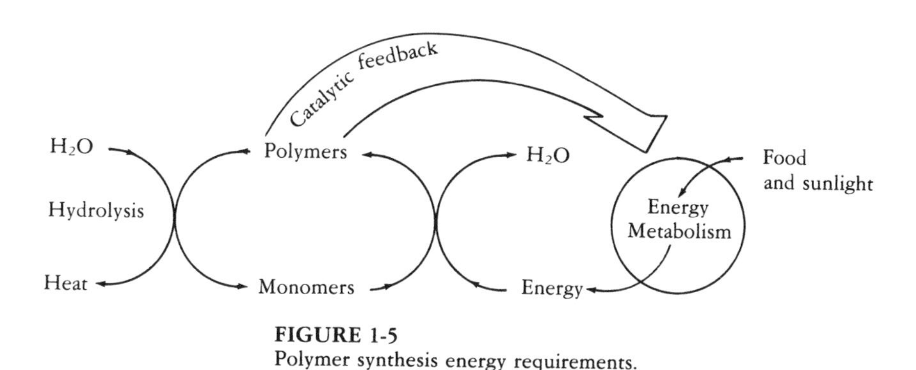


**FIGURE 1-5**
Polymer synthesis energy requirements.

ergy requirement exists because biological polymers are surrounded by water, which tends to degrade them by hydrolysis. Organisms acquire this energy through the catalytic activity of enzymes, special proteins which the organisms manufacture from instructions coded in genes (DNA). This is the essence of the uroboros problem: to make polymers, polymers are needed. Alternatively, this may be expressed: to make the energy needed for synthesis of polymers, energy for synthesis is needed. The transition from monomers to polymers poses the first real difficulty.

How did this uroboros puzzle get solved on the primitive earth? The next section clarifies the questions; the next chapter attacks the problem.

### Summary of Monomer → Polymer Energy Requirement

The relationships between polymer production and energy production are schematized in Figure 1-5.

## 1-6. THE UROBOROS PUZZLE

Stable small molecules and monomers can arise from the primordial dozen elements, as previous discussions have shown. But the setting contemplated had abundant water, partly because the large negative free energy of formation is very favorable for $H_2O$ to form from the elements. Formation of siliceous crustal rock is similarly favored.

### Unfavorable Thermodynamics

A watery environment (such as existed on the primitive earth) is not a hospitable place for linking monomers by dehydration condensations because the reverse process, degradation by hydrolysis, tends to break the resulting polymer into its constituent oligomers and monomers. Indeed, the hydrolysis of polymers is a mechanistically simple, thermodynamically favorable—but fortunately, slow—process.

Therefore, in the presence of abundant $H_2O$, the formation of polymers is thermodynamically unfavorable. When experimenters put polymers, such as proteins and polynucleotides, in water, eventually (although not rapidly) the polymers hydrolyze (break up). And when they put amino acids and nucleoside-monophosphates into water, these molecules do not polymerize.

In general, peptide formation in aqueous milieu requires an increase in free energy of 2 or 3 kcal/mol for each bond formed, and phosphodiester formation requires an increase of about 5 kcal/mol per bond. Glycosides require an intermediate amount of energy per bond. In all cases, formation of dehydration linkages in aqueous milieu requires an increase in free energy, while hydrolysis of dehydration linkages decreases the free energy by 2–5 kcal/mol. That is, hydrolysis is spontaneous, while polymerization is thermodynamically inhibited, although the rate of hydrolysis varies over many orders of magnitude. These conflicting tendencies occur in a ratio determined by Boltzmann's formula $\exp[-\Delta G/RT]$, in which $\Delta G$ measures the free energy change from reactants to products. For a peptide bond with $\Delta G = 3$ kcal/mol, this formula yields a factor of $\exp(-5) = 0.007$ in favor of peptide over monomer. The exponential character of the criterion implies that multiple dehydration linkages are exponentially inhibited—that is, very strongly. For $n$ linkages, the factor is $\exp[-5n]$; for example, for $n = 20$, $\exp[-100] = 3.7 \times 10^{-44}$. Consequently, organisms must use complex strategies to circumvent this fundamental difficulty. (And any solution to the uroboros puzzle must develop one.)

### A Dead-End Solution

One solution might be to use the diversity of the settings on the young planet. Both aqueous zones and hot dry zones must have existed. A hot-dry zone has no $H_2O$ to inhibit and reverse the dehydrations; moreover, every time a dehydration linkage occurs, the liberated $H_2O$ quickly evaporates. Unfortunately, these hot-dry conditions tend to thermally degrade (pyrolyze) the monomers. Moreover, the early atmosphere contained no ozone to shield the surface from the sun's UV radiation, which also degrades the monomers.

Building polynucleotides is even more difficult than building proteins because their monomers (nucleotides) are mixed oligomers and as such are subject to hydrolysis; thus they are unavailable for polymerization. Both polynucleotides and mononucleotides are thermodynamically unfavorable in aqueous milieu. Coenzymes with dehydration linkages also would hydrolyze.

### Activation Energy, an Organism's Solution

A solution to this problem must exist because organisms are largely composed of $H_2O$, as shown in Table 1-1. How do modern organisms produce the polymers and mixed oligomers so essential to the living state? The answer is that they use energy flow to overcome the thermodynamic disadvantage. This answer sounds simple, but its implementation by an organism is not.

*Organisms use energy to activate monomers to a state in which polymerization occurs spontaneously.* They convert this energy into a form suitable for monomer activation, and they catalyze and control the polymerization of *activated monomers*. The polymers that form are still subject to hydrolysis; but as long as the rate of polymerization is fast enough, a nontrivial concentration of polymers can build up. (Chapter 2 describes how this is done.)

The characteristic feature that will emerge from an account of energy metabolism and the polymerization of activated monomers is that *the products of these processes, polymers, are themselves essential components of the molecular apparatus by which polymers are produced.* In this sense, polymers are self-begetting: They are a molecular manifestation of an uroboros and the uroboros puzzle becomes the problem of initiating a self-begetting polymer synthesis mechanism.

Previous discussion of the synthesis of the elements and the subsequent manufacture of small molecules and monomers invoked a variety of plausible energy fluxes. However, formation of dehydration condensations, which is necessary for making polymers from monomers, restricts the mechanism of energy flow; the next chapter discovers that organisms use a remarkably versatile chemical form of energy, based on the relatively rare element, phosphorus, for this purpose.

## REFERENCES


**Section 1-1**

Fowler, W. A., *Nuclear Astrophysics,* American Philosophical Society, Philadelphia, Pa., 1967.

Weinberg, S., *The First Three Minutes: A Modern View of the Origin of the Universe,* Basic Books, New York, 1977.

**Section 1-2**

Calvin, M., *Chemical Evolution,* Oxford University Press, New York, 1969.

Dickerson, R. E., *Molecular Thermodynamics,* W. A. Benjamin, New York, 1969.

Fox, S. W., and K. Dose, *Molecular Evolution and the Origin of Life,* W. H. Freeman, San Francisco, 1972.

Miller, S. L., and L. E. Orgel, *The Origins of Life on the Earth,* Prentice-Hall, Englewood Cliffs, N.J., 1974.

Needham, A. E., *The Uniqueness of Biological Materials,* Pergamon Press, Elmsford, N.Y., 1965.

Walker, J. C. G., *Evolution of the Atmosphere,* Macmillan, New York, 1977.

**Section 1-3**

Calvin, M., *Chemical Evolution,* Oxford University Press, New York, 1969.

Dickerson, R. E., *Molecular Thermodynamics,* Benjamin, New York, 1969.

Fox, S. W., and K. Dose, *Molecular Evolution and the Origins of Life,* W. H. Freeman, San Francisco, 1972.
Oro, J., *Nature,* 190 (1961): 389; 191 (1961): 1193.
Oro, J., and A. P. Kimball, *Arch. Biochem. Biophys.* 94 (1961): 217; 96 (1961): 293.
Miller, S. L., *Science* 117 (1953), 528.
Harada, K., and S. W. Fox, *Nature* 201 (1964): 335.
Stryer, L., *Biochemistry,* 3d ed., W. H. Freeman, New York, 1988.

# CHAPTER 2

# *The Primitive Uroboros*

Two basic ideas lie behind the model of a primitive uroboros that I shall describe in this chapter. The first idea is prevalent in most of the literature on biological origins: Start with something simple with relatively few parts and let it evolve into something complex. The second idea is that the initiation and evolution of an uroboros were driven by some abundant natural source of energy.

Section 2-1 looks at the chemistries that were available on the prebiotic earth and at the special molecules and energy couplings that were needed for life to begin. Several incomplete scenarios provide insight into the problems of working out the steps to a self-begetting and self-maintaining living system. Section 2-2 explores the ways used by contemporary organisms to manufacture polymers and generate phosphate bond energy. In Section 2-3, I present a plausible sequence of events by which a simple, primitive uroboros could have begun and suggest how it could have evolved into greater complexity. A thermodynamic analysis of the proposed sequence is the subject of Section 2-4.

Discussions in this chapter are strongly biochemical in flavor and detail, though emphasis is on concepts. Readers who lack the appropriate background may find the references and glossary useful. Mathematically minded readers may find it fruitful to move on to Chapter 4 and read Chapters 2 and 3 later.

## 2-1. ENERGY-DRIVEN POLYMERIZATION

The beginning of polymerization on the primitive earth required that chemical reactions among some very special molecules be driven by a new form of energy that was based on phosphorus.

### The Chemistry Available on the Prebiotic Earth

Several kinds of chemistry were occurring on the prebiotic earth, including acid-base reactions, in which protons transfer from one type of molecule to another, and oxidation-reduction reactions, in which electrons transfer from one molecule to another. During some oxidation-reduction reactions proton exchanges also take place. The proton-transfer and electron-transfer reactions began early on the primitive earth, since they derived from geophysical energy flows, and no doubt played prominent roles in prebiotic energy transfer.

Another type of chemistry, based on phosphorus, just as assuredly played a special role. Ultimately, phosphorus in the form of phosphate bond energy enabled geophysical oxidation-reduction energy to flow through the organic matter that existed on the primitive earth. *Through conversion into phosphate bond energy, oxidation-reduction energy is capable of driving the polymerization of monomers.* Contemporary organisms use the phosphate bond energy in adenosine triphosphate (ATP), a nucleoside derivative, to drive polymerization. The usable energy in ATP is associated with the pyrophosphate moiety, the anhydride linkage between two phosphates in ATP or—in the simpler case—in pyrophosphate:

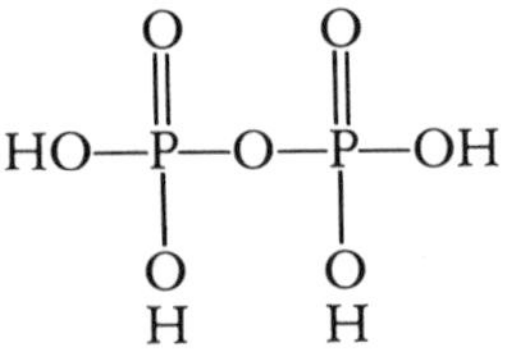

```
       O     O
       ‖     ‖
  HO—P—O—P—OH
       |     |
       O     O
       H     H
```

Pyrophosphate

See Figure 1-4 to compare this structure with that of ATP. Energy is released when the pyrophosphate is split—for example, by hydrolysis, which releases over 7 kcal of free energy per mole as heat.

### The Special Molecules Needed for Polymerization

The exact energy-flow path that caused monomers to polymerize for the first time on the primitive earth is unknown, but plausible hypotheses are possible. Certain kinds of molecules played special roles in the initiation event.

***The Role of Activated Monomers*** Monomers can absorb and sequester the energy released when phosphate bonds are broken. These energy-rich monomers are said to be *activated.* For example, if a phosphate transfers from a pyrophosphate to the carboxyl group on an amino acid, an amino acyl phosphate (a mixed anhydride) results, as shown:

$$H{-}C(R)(N^{+}H_3){-}CO_2^{-} + HO{-}P(=O)(OH){-}O{-}P(=O)(OH){-}OH \rightarrow H{-}C(R)(N^{+}H_3){-}C(=O){-}O{-}P(=O)(O^{-}){-}OH + HO{-}P(=O)(OH){-}OH$$

Amino acid — Pyrophosphate — Amino acyl phosphate (Activated monomer) — Phosphate

This activated monomer can take part in one of two competing reactions: hydrolysis and dimerization. Hydrolysis degrades it, releasing the energy as heat (about 7 kcal/mol):

$$H{-}C(R)(N^{+}H_3){-}C(=O){-}O{-}P(=O)(OH){-}OH + H_2O \rightarrow H{-}C(R)(N^{+}H_3){-}C(=O){-}OH + HO{-}P(=O)(OH){-}OH + \text{heat}$$

Dimerization uses the monomer's activation energy to form a peptide bond between it and another monomer:

$$\mathrm{H{-}\underset{N^+H_3}{\overset{R_1}{C}}{-}\overset{O}{\overset{\|}{C}}{-}O{-}\overset{O}{\overset{\|}{P}}(OH){-}OH + H{-}\underset{N^+H_3}{\overset{R_2}{C}}{-}CO_2^- \rightarrow H{-}\underset{N^+H_3}{\overset{R_1}{C}}{-}\overset{O}{\overset{\|}{C}}{-}\underset{H}{N}{-}\underset{H}{\overset{R_2}{C}}{-}CO_2^-}$$

$$\mathrm{+\ HO{-}\overset{O}{\overset{\|}{P}}(OH){-}OH}$$

If the activation-to-dimerization steps are rapid enough, dimerization will dominate the hydrolysis of the activated monomers and the product dipeptide and lead to a buildup of dipeptide molecules. These molecules can then grow longer by further additions of activated monomers.

To emphasize their dynamic quality, several processes involving activated monomers are depicted as follows:

$P \sim P + W \rightarrow P + P + \text{heat}$ (Hydrolysis of pyrophosphate)

$P \sim P + A_i \rightarrow A_i \sim P + P$ (Activation of amino acid)

$A_i \sim P + W \rightarrow A_i + P + \text{heat}$ (Hydrolysis of activated amino acid)

$A_i \sim P + A_j \rightarrow A_i{-}A_j + P$ (Peptide bond formation)

$A_i \sim P + A_j \sim P \rightarrow A_i{-}A_j \sim P + P$ (Activated peptide formation)

where A denotes an amino acid, $i$ an integer index, P a phosphate group, W a water molecule, the symbol $\sim$ an energy-rich phosphate bond, and the symbol — a typical covalent chemical bond with less free energy content.

The equations show that these reactions are not aided by any catalytic agent. In addition, they are virtually irreversibile.* The peptide bond between $A_i$ and $A_j$ has a $\Delta G$ (free energy content relative to hydrolysis) of only 4 to 5 $\times\ RT_{298}$ kcal/mol. This means that $A_i{-}A_j$ will hydrolyze and release heat, but the bond energy is otherwise not useful:

$$A_i{-}A_j + W \rightarrow A_i + A_j + \text{heat} \quad \text{(Peptide hydrolysis)}$$

* This is only an approximation, albeit a good one when the reaction involves a free energy change large compared with the scale set by the thermal energy: $RT_{298} = 0.59$ kcal/mol. The energy-rich phosphate bonds in phosphoanhydride and amino acyl phosphate groups have free energy of hydrolysis differences of about $10 \times RT_{298}$. This makes hydrolysis virtually irreversible ($e^{-10} \cong 5 \times 10^{-5}$).

The dynamics of this system, stimulated by an external energy flux that generates $P \sim P$, distinguishes it from a system in thermal equilibrium. In thermal equilibrium, without an energy flux, none of the reactants $P \sim P$, $A_i \sim P$, or $A_i—A_j$ could occur because each participant would have hydrolyzed to its monomeric form. However, an energy flux that continually generates $P \sim P$ will drive this system forward, producing populations of $A_i—A_j$. Moreover, if the system includes polymer-lengthening reactions, then a distribution of polymer lengths will also result, as follows:

$$A_i—A_j—A_k \sim P + A_l \sim P \rightarrow A_i—A_j—A_k—A_l \sim P + P$$

(Peptide elongation)

As previously stated, the rates of competing reactions determine the outcome. If hydrolysis dominates, no net synthesis of polymers takes place; if activation and polymerization are fast enough, synthesis prevails, and a distribution of lengths and sequences results. The hydrolysis rate for a single peptide bond is only $10^{-6}$ to $10^{-7}$ per second, that is, roughly one a month. However, experiments (Fox and Dose, 1977) confirm that hydrolysis dominates *unassisted* activation and polymerization.

***The Role of Catalytic Peptides*** Assisted activation is the key to producing biologically interesting *macromolecules,* which can have so-called emergent properties. *Emergent properties* of systems can be predicted only with great difficulty from knowledge of the properties of a system's components. Thus the properties of macromolecules are difficult to predict from the behaviors of monomers or oligomers. One emergent property is that some macromolecules, called *enzymes,* act as biological catalysts. A modest catalytic activity can easily dominate a hydrolysis rate of one peptide bond per month. A turnover of $10^{-4}$ per second is 100 times faster than hydrolysis though it is $10^{14}$ times slower than the best contemporary enzyme. Carbonic anhydrase holds the record for the highest rate of turnover of substrate molecules: $6 \times 10^{10}$ per second.

## Incomplete Constructs

The following hypothetical and incomplete constructs cannot initiate a molecular uroboros, but they demonstrate the problems involved.

***Polypeptide Cross-Catalysis (Hypothetical)*** Assume that three particular pentapeptides (sequences of five amino acids) exist:

$$A_{a1}—A_{a2}—A_{a3}—A_{a4}—A_{a5}$$

$$A_{b1}—A_{b2}—A_{b3}—A_{b4}—A_{b5}$$

$$A_{c1}—A_{c2}—A_{c3}—A_{c4}—A_{c5}$$

which are abbreviated $E_a$, $E_b$, and $E_c$, respectively. Further, suppose that the pentapeptides exhibit catalytic properties: $E_a$ catalyzes the formation of $P \sim P$ from a hypothetical energy flux, $E_b$ catalyzes activation of amino acid $A_i$ by $P \sim P$, and $E_c$ catalyzes polymerization of $A_i \sim P$ and $A_j \sim P$:

$$\text{energy flux} + 2P \xrightarrow[E_a]{} P \sim P$$

$$A_i + P \sim P \xrightarrow[E_b]{} A_i \sim P + P$$

$$A_i \sim P + A_j \sim P \xrightarrow[E_c]{} A_i\text{—}A_j \sim P + P$$

Assume that these three catalytic functions proceed fast enough so that their rates prevail over the hydrolysis rate. That is, although these catalytic pentapeptides are also subject to hydrolysis, they possess long half-lives. If they were essential components, they would appear, regenerated, among the products of the polymerizations. This is part of the uroboros requirement.

***Protein Synthesis (Incomplete Construct)*** Mixtures of amino acids, heated to dryness, yield short chains of amino acids called thermal proteins, or *proteinoids*. Experiments (Fox and Dose, 1977) have shown that the lengths and the amino acid sequences of proteinoids are not random but are determined largely by the composition of the amino acid mixture and other reaction conditions, and that the yield is aided by the addition of phosphates. Table 2-1 shows the constituents of amino acid mixtures in various environments.

About the time life began, this process could have been active on the earth's crust. Pools and lakes of liquid water containing dissolved small molecules, including amino acids, had accumulated in shallow depressions in the lava. When the water evaporated or splashed onto the surrounding rock, heat energy in the rock would have reacted the amino acids to form proteinoids; then rain refilled the pools or tides washed the proteinoids back into the water.

If some of the proteinoids possessed catalytic functions such as those of $E_a$, $E_b$, and $E_c$, then polymer synthesis would begin. Such a self-begetting system would also be self-maintaining: the primitive uroboros.

Figure 2-1, an elaboration of Figure 1-5, represents this incomplete construct of protein synthesis.

***Polynucleotide Synthesis (Incomplete Construct)*** An energy-driven transition from nucleotide monomers to polynucleotide polymers offers several attractive features that are not possible with proteins.* The structure of these oligonucleotide molecules supports the existence of *base-pairing* mechanism

* The reader should consult the glossary or a biochemistry textbook, such as that by L. Stryer, for definitions of terms.

**TABLE 2-1**
***Relative Abundances of Amino Acids in Various Environments****

| Cosmic Sources | | |
|---|---|---|
| *Moon* | *Meteorites* | *Hot terrestrial lava* |
| Glycine | Glycine | Glycine |
| Alanine | Alanine | Alanine |
| Glutamic acid | Glutamic acid | Glutamic acid |
| Aspartic acid | Aspartic acid | Aspartic acid |
| Serine | Valine | Serine |
| Threonine | Proline | Threonine |
| | | Valine |
| | | Isoleucine |
| | | Leucine |

| Laboratory Simulations | |
|---|---|
| *Reducing atmosphere and electric discharge* | *Heating* |
| Glycine | Glycine |
| Alanine | Alanine |
| Glutamic acid | Glutamic acid |
| Aspartic acid | Aspartic acid |
| | Serine |
| | Valine |
| | Proline |

* The relative abundance decreases downward on these lists. The crucial reaction condition for successful synthesis of a substantial yield of polypeptides is a relative abundance of aspartic and glutamic acids; otherwise, heating amino acid mixtures yields dark sticky tar. Compositional analyses of hundreds of proteins from organisms show relative abundances of these two amino acids. Indeed, in animals, glutamic acid is in greatest abundance.

for lining up activated nucleotide monomers on an oligonucleotide template so that the product contains a sequence homologous to the sequence used as the template. Moreover, the double-stranded form is literally replicated by the base-pairing mechanism of polymerization. This mechanism reproduces both the structure and the sequence: For the single-stranded form, the first

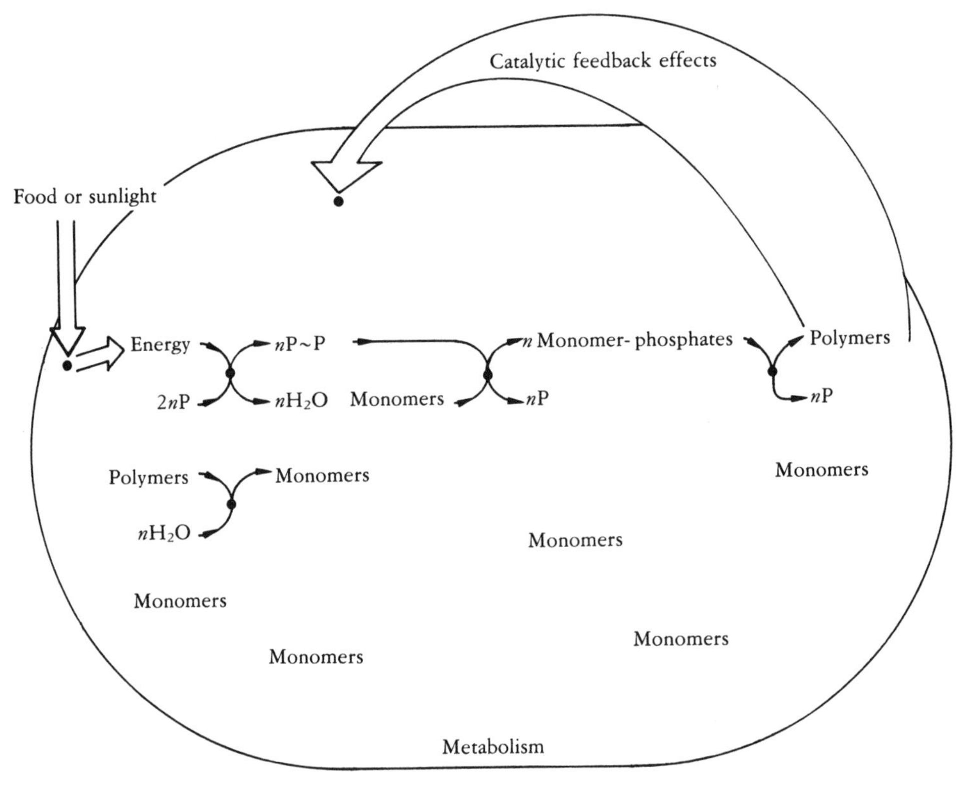


**FIGURE 2-1**
Energy requirements for polymer synthesis in proteinoid microsphere. Large dots denote proteinoid's catalytic feedbacks on metabolism and on monomer activation and polymerization.

base-pairing has a complementary sequence, and the complement of the complement has the original sequence, as shown in the following equations:

$$N_i + 2P \sim P \rightarrow N_i \sim P \sim P + 2P \quad \text{(Nucleoside activation)}$$

$$N_i \sim P \sim P + 2W \rightarrow N_i + 2P + \text{heat} \quad \text{(Hydrolysis of activating phosphates)}$$

$$\overline{N}_i \sim P \sim P + \overline{N}_j \sim P \sim P \rightarrow \overline{N}_i\text{—}\overline{N}_j \sim P \sim P + P \sim P \quad \text{(Template activity)}$$

$$N_i\text{—}N_j$$

where $N_i$ denotes a nucleoside monophosphate which already contains one phosphate not involved in activation, $P \sim P$ a pyrophosphate, $N_i \sim P \sim P$ an activated nucleoside triphosphate, W denotes water, and $\overline{N}_i$ the base-pairing complement to $N_i$. The last reaction shows the base-pairing template behavior of $N_i$—$N_j$.

Without a template, activated nucleotides do not spontaneously polymerize to any appreciable extent. But where could a template have come from initially? Another severe problem exists: Even in experiments with polynucleotide templates and activated nucleotides, polymerization is difficult to achieve because the activated monomers tend to hydrolyze faster than they polymerize. In contemporary organisms, the slow reaction rate of nucleotides is overcome by enzymes called polymerases; and the hydrolysis of activated monomers is overcome by series of reactions, also catalyzed by enzymes, that continuously regenerate the monomers. In addition to these problems, the coupling of any environmental energy flux to $\sim P \sim P$ production, which is required for polynucleotide synthesis, also requires proteins. How could an uroboros have been started?

The pathways that organisms use to regenerate the critical monomers help in understanding the problems of polynucleotide synthesis and uroboros initiation. Organisms do not generate purine mononucleotides simply by linking a ribose molecule, a phosphate ion, and a purine base; they begin with ribose-monophosphate (in a linkage that is not energyrich). Through a succession of reactions—involving pieces of aspartic acid and glutamine (amino acids), formic acid, and carbamyl phosphate—organisms synthesize the purine in place, attached to the sugar. These reactions require energy. To synthesize purines, an organism uses no fewer than five high-energy phosphate bonds (from five ATPs) to activate intermediates in the pathway. Enzymes catalyze every step. These observations imply that a protein-generating system must have been in place before polynucleotides became abundant. They also emphasize that if phosphate bond energy was plentiful, then an amino acid-to-protein system could have been driven in the direction of these later phosphate-dependent processes. It is my view that modern polynucleotides are a manifestation of a continuing stage of energy transduction development in which abundant phosphate bond energy dominated, and still dominates, the evolution of organic matter. This view is elaborated later with respect to metabolic control, development mechanisms, and major changes in phylogeny.

Nevertheless, the base-pairing mechanism of nucleotide polymerization has somehow given rise to the genetically controlled, sequential synthesis of specified proteins. Obviously a coupling of these two critical types of polymerization (polynucleotides and proteins) has evolved. Thus, even though it is not likely that they arose without preexisting proteins,* polynucleotides will play a key role in the initiation of uroboros.

## The Original Energy Coupling

Neither of the polymerization schemes just described will work without a source of phosphate free energy, $P \sim P$. Since polymerization must be energy-driven, somehow the necessary energy coupling must have taken place on the

* The recent suggestion that RNA served as its own synthetase (Cech, 1986) is not relevant here because it requires a very large RNA as catalyst (300 to 400 bases).

primitive earth. The need here is to find a simple mechanism because the difficulty that polymerization presents is so great that it would be insurmountable if the initial energy coupling were also difficult to achieve.

Therefore, I envisage the initiation of life in conditions of abundant, geophysically generated oxidation-reduction energy that was created during the iron catastrophe. How this oxidation-reduction energy was first converted into phosphate-bond energy is not known precisely, but several hypotheses exist. One possible mechanism uses iron-sulfur compounds. Phosphate can bond to sulfur in such a compound by a linkage that becomes energy-rich when it is oxidized by iron(III). Phosphorolysis then transfers this activated phosphate to free phosphate, yielding pyrophosphate. Since the iron is reoxidized geophysically (probably by a ferric iron compound during the iron catastrophe), the iron-sulfur compound can act cyclically.

Another possible mechanism involves the compound carbamyl phosphate, which forms spontaneously in aqueous mixtures of phosphate and cyanate, $CNO^-$. Cyanate has a positive free energy of formation relative to the elements carbon and nitrogen, but it forms easily when energy (electrical discharges and UV light) flows through mixtures of gases containing carbon- and nitrogen-rich molecules (such as $CO_2$ and $NH_3$). Carbamyl phosphate is energy-rich and could be a source of phosphate bond energy (Lipmann, 1965) for the synthesis of the much more versatile phosphate carrier, pyrophosphate, as follows:

$$HN{=}C{=}O + HO{-}\overset{\overset{O}{\|}}{\underset{\underset{O^-}{|}}{P}}{-}O^- \rightarrow H_2N{-}\overset{\overset{O}{\|}}{C}{-}O{-}\overset{\overset{O}{\|}}{\underset{\underset{O^-}{|}}{P}}{-}O^-$$

(Carbamyl phosphate formation)

$$H_2N{-}\overset{\overset{O}{\|}}{C}{-}O{-}\overset{\overset{O}{\|}}{\underset{\underset{O^-}{|}}{P}}{-}O^- + HO{-}\overset{\overset{O}{\|}}{\underset{\underset{O^-}{|}}{P}}{-}O^- \rightarrow HO{-}\overset{\overset{O}{\|}}{\underset{\underset{O^-}{|}}{P}}{-}O{-}\overset{\overset{O}{\|}}{\underset{\underset{O^-}{|}}{P}}{-}O^-$$

$$+ \underset{\text{(Carbamate)}}{H_2N{-}\overset{\overset{O}{\|}}{C}{-}O^-}$$

(Pyrophosphate formation)

## Transition from Heterotrophic to Autotrophic Polymerization

In the most primitive stages of the development of polymerization of monomers, polymerization required ready-made molecules and thus was *heterotrophic.* The monomers were available as by-products of geophysical energy fluxes through the chemical elements. Amino acids were easily made and were used directly. As stressed before, mononucleotides were not so available. Thus only a simple amino acid-polymer heterotroph could have existed. Later, coupled protein-polynucleotide synthesis emerged. Much later, (a half billion years), life developed photosynthesis and the ability to synthesize all needed components, including amino acids and mononucleotides; that is, polymerization became completely *autotrophic.* Autotrophic function, pressured by depletion of ready-made monomers by heterotrophs, must have developed slowly during this half billion years as the metabolic pathways responsible for the synthesis of components and monomers evolved.

## 2-2. ENERGY METABOLISM AND POLYMER SYNTHESIS IN CONTEMPORARY ORGANISMS

Surveying the mechanisms used by contemporary organisms to generate phosphate bond energy and to manufacture polymers provides clues about the transition from no-life to life and suggests the detailed components and mechanisms that the primitive uroboros must have to be recognized as living.

The cell uses proteins in many ways, and most of the types of proteins that are used are enzymes for catalyzing many reactions, including those in energy metabolism and in the synthesis of monomers, proteins, and polynucleotides. Cells use polynucleotides primarily for governing protein synthesis. The base-pairing behavior of nucleotides tightly controls the amino acid sequences of proteins and also guarantees sequence fidelity during polynucleotide replication for organism reproduction.

### Pathways of Energy Metabolism

An energy metabolism pathway is a series of reactions that makes energy available to cells. The principal pathways used by organisms are *glycolysis,* the pentose phosphate pathway (or hexose monophosphate shunt), and the *electron transport chain*; the last occurs in two forms, one for photosynthetic cells and another for aerobic cells. In aerobic cells, carbohydrate oxidation and electron transport are connected by the *tricarboxylic acid cycle* (Kreb's cycle) which extracts electrons from glycolysis and places them in intermediate molecules, which then feed them to the electron transport chain.

The details in Figures 2-2, 2-3, and 2-4 serve to emphasize the central importance of energy processing. These pathways are complex, have many steps, and require ~P and enzymes.

***Electron Transport Chain*** In each of the preceding sequences, formation of NADH or $FADH_2$ captures some carbohydrate bond energy that the electron transport chain eventually uses to make ATP. The electron transport chain is a sequence of reactions that involve iron-sulfur proteins, quinone, and *cytochromes.* Finally, the electrons and two protons reduce molecular oxygen

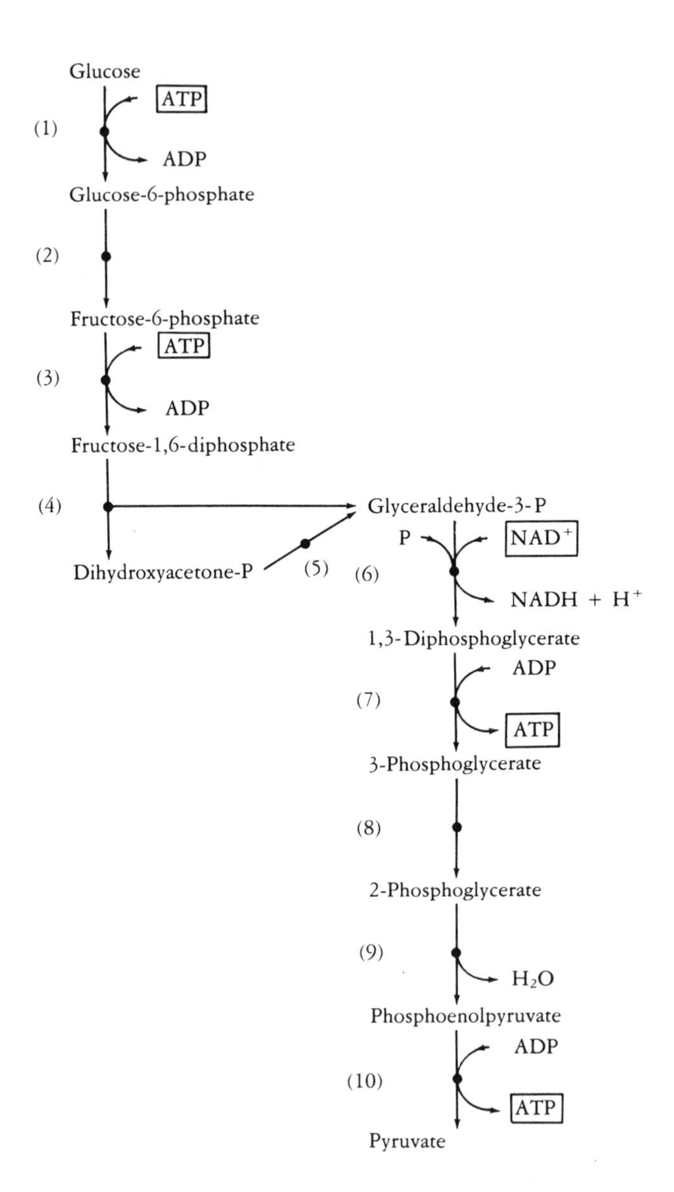


**FIGURE 2-2**
Glycolysis: glucose to pyruvate. During glycolysis, one glucose molecule eventually becomes two molecules of pyruvate and two molecules of water. In this process, an investment of two molecules of adenosine triphosphate (ATP), an energy-storing phosphate molecule used extensively in biology, is necessary to activate the pathway, which eventually yields four ATPs [steps (6)–(10) occur twice per cycle] for a net gain of two. These high-energy phosphate molecules carry away some of the free energy of the carbohydrate. Glycolysis also produces two molecules of the reduced coenzyme NADH which carry the electrons yielded by oxidation to the electron transport chain. The $NAD^+$ step, (6), is an *oxidative phosphorylation,* accomplished through the intermediate carrier of energy-rich phosphate, 1,3-diphosphoglycerate. The total energy balance sheet for glycolysis is

$$\text{glucose } (CH_2O)_6 + 2NAD^+ + 2ADP + 2P \rightarrow 2 \text{ pyruvate} + 2H_2O + 2ATP + 2NADH + 2H^+$$

This process proceeds as written, in the direction of decreasing free energy overall as an organism uses glucose. Glycolysis is called *primitive* because it does not require $O_2$. Steps (6) and (7) will be of interest in Section 2-3.

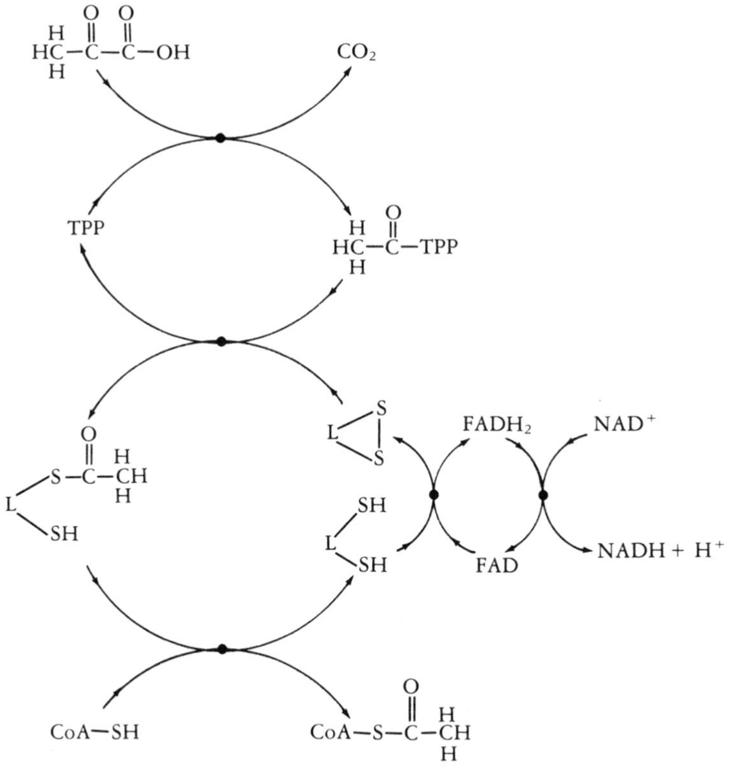

**FIGURE 2-3**
Pyruvate to acetyl—CoA. A group of coupled reactions oxidizes and decarboxylates the pyruvate from glycolysis. These reactions simultaneously involve several coenzymes and enzymes. The coenzymes are FAD, LSS, NAD, TPP, and CoA (see Table 1-4). The product of this group of reactions is acetyl—CoA, an energy-rich form of acetate that is linked to the coenzyme CoA—SH through a dehydration linkage called a thioester. The total balance sheet can be written as

$$\text{pyruvate} + NAD^+ + \text{CoA—SH} \rightarrow \text{acetyl—CoA} + NADH + H^+ + CO_2$$

which also proceeds in the direction of decreasing free energy.

to water. Some of the steps of this chain are purely electron-transfer oxidation-reduction reactions, whereas others involve protons. Those steps that include proton transfers move protons across the cell membrane, creating an electrochemical potential energy across the membrane. In a recently elucidated process called chemiosmosis, this membrane potential drives ADP and P to ATP, thereby completing the transduction of oxidation free energy into phosphate bond energy. These events are depicted in Figure 2-5.

The iron-sulphur proteins in the electron transport chain are thought to be very ancient. They contain iron that is covalently linked to sulfurs in a tetrahedral arrangement similar to that found in crystals of the mineral chalcopyrite. A protein binds both NADH and FAD and begins the electron transport chain

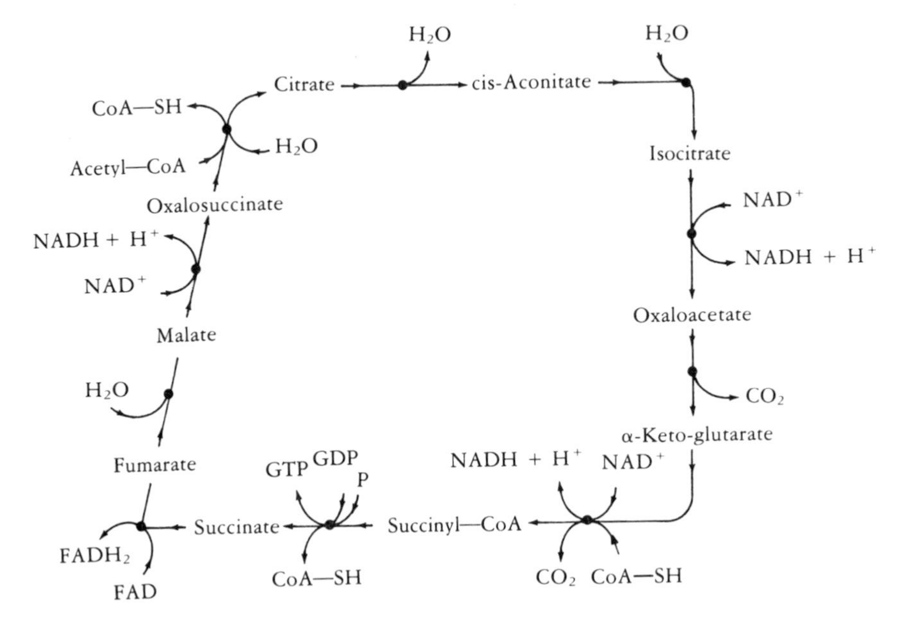


**FIGURE 2-4**
Tricarboxylic acid cycle. One use of acetyl—CoA is to be incorporated into the tricarboxylic acid cycle. In this pathway, acetyl—CoA reacts with oxaloacetate to form citrate. Group rearrangements, oxidations, and decarboxylations ensue, as well as a phosphorylation of guanine diphosphate (GDP) to guanine triphosphate (GTP), a close relative of ATP but with the base guanine rather than adenine. The overall balance sheet for this cycle is

$$\text{CoA—SH} + \text{acetyl—CoA} + 3\text{NAD}^+ + \text{FAD} + \text{GDP} + \text{P} \rightarrow$$
$$2\text{CoA—SH} + 3\text{NADH} + 3\text{H}^+ + \text{FADH}_2 + 2\text{H}_2\text{O} + \text{GTP} + 2\text{CO}_2$$

This process eventually frees the two carbons from the acetate, in their fully oxidized form as $2CO_2$. The electrons carried away by 3NADH and FADH retain much of the original free energy of the acetate, and another portion of the free energy is in the phosphate-bond potential of GTP. Oxaloacetate reforms, and with the addition of more acetyl—CoA, the cycle continues.

by extracting electrons from NADH, transferring them to FAD, and forming $FADH_2$, which then transfers them to one of these iron-sulfur proteins.* The cytochromes contain heme iron (iron complexed to nitrogen atoms in a porphyrin ring) which is linked to the protein part of the molecule by sulfur bridges (bonds). The several cytochromes are distinguishable by their protein components.

The energy metabolism pathways just described depend on a large number of enzymes whose synthesis is driven by the energy generated by the pathways.*

* Electron transport, which depends on oxygen, is more complex (and more recent) than glycolysis. In bacteria, it occurs in the cell membrane, correlated with an intricate spatial arrangement of the membrane proteins, whereas glycolysis occurs within the gel-like matrix of the cell interior, without oxygen or special arrangement of enzymes. Most of the cell's total carbohydrate energy is harvested by electron transport (about 90 percent) rather than by glycolysis.

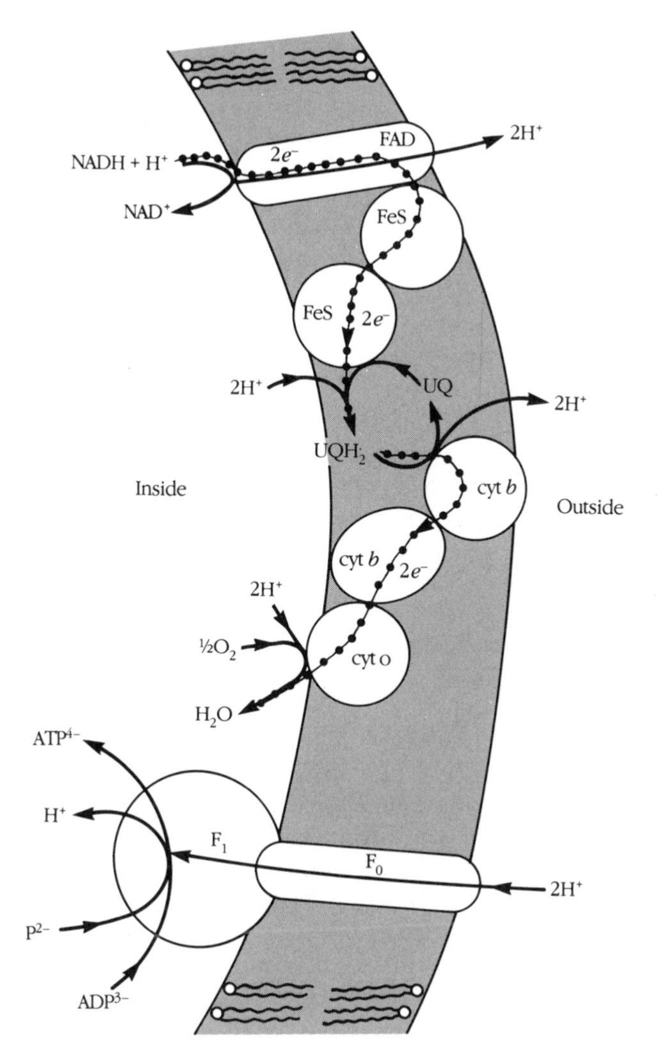


**FIGURE 2-5**
Electron transport in bacterial membrane. Solid lines denote proton movements; dotted lines, electron movements; UQ, ubiquinone; cyt, cytochrome; FeS, iron-sulfur protein; and $F_1$ and $F_0$, protein complexes comprising the ATP synthesizing enzyme.

## Genetic Control of Protein Synthesis

The coded instructions that specify the amino acid sequences of all the proteins needed by an organism reside in its genes. The process of protein synthesis as it relates to the beginning of life appears in this section; the reader should refer to books on biochemistry or molecular biology for greater detail (for example, Lewin, 1986; Stryer, 1981; and Watson, 1975).

The genes are the active portions of enormous double-stranded molecules of the polynucleotide, deoxyribonucleic acid (DNA), known as chromosomes.

Each strand consists of a sugar and phosphate backbone, to which are attached the nucleic acid bases. Connecting the two spiraling strands of DNA are numerous hydrogen bonds, which form between the pairs of bases: Adenine (A) and thymine (T), guanine (G) and cytosine (C). When sex cells are made, enzymes replicate the strands of DNA in such a way that every offspring inherits a full complement of parental genes. Before replication can begin, enzymes must separate the entwined strands and line up energy-activated mononucleotides along the separated strands according to base-pairing rules (A-T and G-C), and phosphodiester bonds must form between adjacent activated nucleotides. If these events could occur spontaneously, DNA would indeed be self-replicating, provided the activation energy were available. However, replication requires the aid of numerous enzymes. For instance, in *E. coli* bacteria, initiation of DNA replication requires the so-called unwinding enzymes and RNA polymerase (a large protein). After synthesis of a short stretch of ribonucleic acid (RNA), DNA polymerase III (another large protein) enters the picture. Finally, DNA polymerase I and DNA ligase finish the job, which has used 12 to 20 enzymes.

Constructing a protein from the code in a gene is one of the most complex processes in cell biology. After enzymes unwind and separate the two strands of DNA, the enzyme RNA polymerase transcribes the gene into messenger RNA (mRNA), synthesizing a new chain of nucleotides using the gene as template; so the mRNA sequence is always complementary to the gene's DNA sequence. An earlier and separate transcription has made three ribosomal RNAs (rRNA), which organize into a cellular body called a *ribosome*, the locus of protein synthesis. The ribosome binds the strand of mRNA. Transfer RNAs (tRNA), also transcribed earlier from special genes, contain a sequence of three bases called an anticodon, which by base-pairing, can "read" sequences of three bases (*codons*) on the mRNA. The kinds of bases on the codon and their order constitute the code for one amino acid of the protein being constructed. At least one specific tRNA exists for each amino acid. The ribosome forms a complex with mRNA and tRNA to read the mRNA and catalyze the formation of a peptide bond between the last amino acid on the growing chain and each new amino acyl tRNA that is brought to it.

The energy for protein synthesis is derived from ATP; an enzyme called tRNA synthetase, binds its tRNA, its amino acid, and ATP. First, the ATP bonds to the amino acid, activating it into an amino acyl adenylate. Then the synthetase (which is specific to the tRNA) bonds this activated amino acid to the tRNA. The resulting amino-acyl-tRNA is still sufficiently energy-rich to form a spontaneous peptide bond. In other words, as a ribosome moves along the m-RNA strand, everytime it reads a codon it "calls in" the corresponding tRNA, which positions its amino acid on the growing chain, where the peptide bond anchors it.

With a triplet code and only four bases, there are at most 64 codons for 20 amino acids. Thus the genetic code is redundant, as is shown in Table 2-2. This redundancy is accommodated by 30 to 40 tRNAs and at least 20 tRNA synthetases.

TABLE 2-2
*The Genetic Code*

| | *U* | | *C* | | *A* | | *G* | |
|---|---|---|---|---|---|---|---|---|
| U | UUU | Phe | UCU | Ser | UAU | Tyr | UGU | Cys |
| | UUC | Phe | UCC | Ser | UAC | Tyr | UGC | Cys |
| | UUA | Leu | UCA | Ser | UAA | End | UGA | End |
| | UUG | Leu | UCG | Ser | UAG | End | UGG | Trp |
| C | CUU | Leu | CCU | Pro | CAU | His | CGU | Arg |
| | CUC | Leu | CCC | Pro | CAC | His | CGC | Arg |
| | CUA | Leu | CCA | Pro | CAA | Gln | CGA | Arg |
| | CUG | Leu | CCG | Pro | CAG | Gln | CGG | Arg |
| A | AUU | Ile | ACU | Thr | AAU | Asn | AGU | Ser |
| | AUC | Ile | ACC | Thr | AAC | Asn | AGC | Ser |
| | AUA | Ile | ACA | Thr | AAA | Lys | AGA | Arg |
| | AUG | Met | ACG | Thr | AAG | Lys | AGG | Arg |
| G | GUU | Val | GCU | Ala | GAU | Asp | GGU | Gly |
| | GUC | Val | GCC | Ala | GAC | Asp | GGC | Gly |
| | GUA | Val | GCA | Ala | GAA | Glu | GGA | Gly |
| | GUG | Val | GCG | Ala | GAG | Glu | GGG | Gly |

| *Amino acid* | *Number of codons* |
|---|---|
| Ala | 4 |
| Arg | 6 |
| Asn | 2 |
| Asp | 2 |
| Cys | 2 |
| Gln | 2 |
| Glu | 2 |
| Gly | 4 |
| His | 2 |
| Ile | 3 |
| Leu | 6 |
| Lys | 2 |
| Met | 1 |
| Phe | 2 |
| Pro | 4 |
| Ser | 6 |
| Thr | 4 |
| Trp | 1 |
| Tyr | 2 |
| Val | 4 |
| End | 3 |

The ribosomes contain about 50 proteins, but the ability of the ribosome to coordinate the translation of m-RNA into protein requires the action of still other proteins, called factors. Some factors function only if they bind to and use the energy of GTP, the energy-rich analogue of ATP. The energy of GTP does not activate the formation of the peptide bond between amino acids but instead ensures smooth functioning of the complex of ribosome, mRNA, and tRNAs.

As the preceding sketch of protein synthesis emphasizes, the replication of genes depends on proteins, and the synthesis of proteins depends on genes, RNA, and other proteins. Both processes require activation energy, which is provided by ATP. ATP is generated by the energy metabolism pathways, which depend on enzymes. This is the uroboros puzzle. How could it have begun? At some time, it must have been simpler.

In the contemporary organism, both energy metabolism and biopolymer synthesis are elaborate molecular mechanisms, where energy requirements for polymer synthesis are met by energy metabolism pathways that need numerous catalysts. But these catalysts are the products of polymer synthesis. To ensure sufficient specificity and appropriate quantities of catalysts, the cell's genetic apparatus directs protein synthesis, which requires the synthesis of several types of polynucleotides (DNA and RNAs). And polynucleotide synthesis requires energy and protein catalysts. Together, the processes of energy metabolism, gene-directed protein synthesis, and gene replication constitute a molecular uroboros, but one of considerable complexity. The problem of how to initiate a molecular uroboros can be solved only if the primitive uroboros possesses a much simpler structure and also has the capacity to evolve into contemporary complexity. The next section presents a model for a primitive molecular uroboros.

## 2-3. INITIATION OF AN UROBOROS

Throughout the discussions that follow and lead to the proposed model for the primitive uroboros, the reader should carry a feeling for geological time. Initiation of the relatively simple, primitive state of the uroboros did not happen quickly, even after conditions on the primitive earth were favorable for life. A look at the history of biological change on the geological time scale helps one appreciate the time scale of the earliest stages of evolution were (see Table 2-3).

Though fossils of microorganisms are scarce, and fossils of macromolecules are nonexistent, several clues to early life do exist. About 3.5 billion years ago, bacteria growing in extensive clumps lay down the substrates for stromatolite fossils (Ballard, 1983, p. 44), the oldest of which are in Australia. These bacteria were probably heterotrophs, which did not use photosynthesis

**TABLE 2-3**
*Geological Time Scale for Evolution*

| *Era* | *Period* | *Beginning (years* B.P.*)* | *Significant events* |
|---|---|---|---|
| CENOZOIC | Quarternary | $2 \times 10^6$ | Development of man |
| | Tertiary | $63 \times 10^6$ | Mammals dominate; modern plants spread and diversify |
| MESOZOIC | Cretaceous | $140 \times 10^6$ | Extinction of dinosaurs; mammals advance; flowering plants appear |
| | Jurassic | $200 \times 10^6$ | Primitive mammals, flying reptiles; birds appear dinosaurs dominate; conifers abundant |
| | Triassic | $240 \times 10^6$ | Reptiles dominate; dinosaurs appear; cycads appear |

TABLE 2-3
*Geological Time Scale for Evolution (continued)*

| *Era* | *Period* | *Beginning (years* B.P.*)* | *Significant events* |
|---|---|---|---|
| PALEOZOIC | Permian | $290 \times 10^6$ | Reptiles advance; huge plants decline |
| | Pennsylvanian | $330 \times 10^6$ | Coal forms from undecayed layers of vegetation |
| | Mississippian | $360 \times 10^6$ | Insects abundant; amphibians advance; first reptiles; tree ferns; conifers |
| | Devonian | $410 \times 10^6$ | First trees and forests; amphibians appear; primitive insects, scorpions, and spiders on land |
| | Silurian | $430 \times 10^6$ | Fishes advance; ozone blocks UV; plants spread on land; coral reefs extensive |
| | Ordovician | $500 \times 10^6$ | Fishes (primitive vertebrates); explosion in the variety of sea life (invertebrates); deposits of limestone |
| | Cambrian | $570 \times 10^6$ | Mosses; diversification of invertebrates such as brachiopods, snails, trilobites, sponges |
| PRECAMBRIAN | Proterozoic | $700 \times 10^6$ | First organisms visible to unaided eye; sponges |
| | | $900 \times 10^6$ | Sexual reproduction; lime-secreting algae |
| | | $1600 \times 10^6$ | First aerobes |
| | | $1800 \times 10^6$ | Stromatolites widespread |
| | | $2100 \times 10^6$ | Early red sandstones ($O_2$ in atmosphere) |
| | | $2500 \times 10^6$ | Stromatolites increase |
| | Archean | $2700 \times 10^6$ | Photosynthesis by blue-green algae |
| | | $3500 \times 10^6$ | Bacterial stromatolite fossils |
| | | $3800 \times 10^6$ | Iron catastrophe; origin of oceans; origin of life |
| | | $4500 \times 10^6$ | Accretion of the planet Earth |

or oxygen for energy; later I shall propose how such early organisms may have been energized. At this time, the genetic apparatus for protein synthesis and the rudiments of energy metabolism were evolving. Stromatolites were still in their ascendancy 2.5 billion years ago, even though photosynthesis had evolved (2.7 to 3.0 billions years ago). By 2.1 billion years ago the first red sandstones formed—red because the $O_2$ content of the atmosphere had increased enough (as a result of photosynthesis) to rust the iron in minerals. However, only about 1.6 billion years ago did atmospheric oxygen become plentiful enough to support aerobic life. Sexual reproduction began 900 million years ago, and the first metazoans large enough to be visible arose about 700 million years ago. All these events occurred in the first 3 billion years of life, in the geological era known as the Pre-Cambrian.

All advanced forms of life have evolved during the last half billion years. Every new fossil find changes some detail of the picture, such as which organism first arrived on dry land. But whichever one did, a sufficient buildup of atmospheric $O_2$ occurred first, forming an ozone layer in the upper atmosphere. This ozone layer is essential because it filters out UV radiation from the sun, which would have been deadly to terrestrial life. The ozone layer initially occurred during the Ordovician period, around 400 million years ago, and saturated during the Silurian.

## The Character of Primordial Life

The energy metabolism of the earliest life on earth was probably neither photosynthetic nor aerobic, but was similar to glycolysis or to some segment or analogue of it. As already mentioned, the glycolitic pathway is independent of photosynthesis and the electron transport chain. It is the least sophisticated of the energy pathways, structurally, in that its enzymes function individually and are water soluble, whereas the enzymes of the other two pathways are highly organized within lipid membranes. Nevertheless, glycolysis depends on enzymes, and enzymes require the mechanism of protein synthesis. Relics of primitive energy metabolism do not exist, even in the least sophisticated contemporary pathways. The links with primordial mechanisms are long extinct, and their macromolecular substructures were not fossilized. Thus, although glycolysis is the most primitive contemporary energy pathway, it cannot be called primordial because of its dependence on enzymes. Similarly, the contemporary gene-directed protein biosynthesis machinery does not contain primordial relics. And those links with the primordial also became extinct during the first half-billion years of life.

There are no relics, but there may be clues. Some key enzymes in glycolysis and other pathways have sulfur-containing amino-acid side chains in their active sites; other enzymes use nonheme iron. Pyrophosphate doubles for ATP in some microorganisms (Kulaev and Vagabov, 1983). Perhaps a mechanism for coupling oxidation-reduction energy to phosphate bond energy could have been a system that used sulfur and iron and did not require modern enzymes. The system had to promote polymerizations of amino acids and nucleotides without the complicated interactions of ribosomes, DNA, RNAs, synthetases, polymerases, and protein factors. The system must have been simple (and the organism must have been simple) and must also have had the capability for evolving into contemporary complexity.

### Primordial Energy Transduction

The coupling of oxidation-reduction energy to phosphate bond energy "might have been the first event on the way to life" (Lipmann, 1971). Even after the earth had cooled enough for liquid water to accumulate, much thermal activity was still present, causing some areas of the earth's crust to be wet and some to be hot and dry. The oxidation state of iron in the crust and in other molecular combinations was either Fe(II) or Fe(III) depending on the environment. So iron is one potential source of readily available oxidation-reduction energy. Another is sulfur, which occurs in different oxidation states, such as sulfite and sulfate, each of which has a large negative free energy of formation (Table 1-2). The transduction of this oxidation-reduction energy into a more versatile and usable form for organic synthesis requires its conversion into phosphate bond energy. This conversion, in my hypothesis, could have been driven by geophysical conditions that may have been especially strong during the iron catastrophe.

Looking for clues in contemporary systems suggests two models of the origin of energy transduction: oxidative phosphorylation and chemiosmotic phosphorylation. The mechanism of substrate-level oxidative phosphorylation, such as occurs in glycolysis, may have come earlier than chemiosmotic phosphorylation, which occurs aerobically and photosynthetically. The only other known case of substrate-level phosphorylation occurs in *Thiobacillus* bacteria, a sulfur-oxidizing organism. Both of these oxidative phosphorylation mechanisms involve sulfur. A closer examination might reveal early mechanistic analogues.

In glycolysis, the steps of interest are those that metabolize glyceraldehyde-3-P into 1,3-bisphosphoglycerate, which in turn phosphorylates ADP to ATP (Figure 2-6); the phosphate in the 3-position is not involved in energy coupling. Step (1): The enzyme works through a sulfur atom on a cysteine residue at the active site to form a thiohemiacetal enzyme intermediate. Step (2): $NAD^+$ oxidizes the 1-carbon, forming a thioester, that is, an aldehyde oxidizes to a carboxylate. Step (3): Inorganic phosphate phosphorolyzes the thioester, forming 1,3-bisphosphoglycerate and recycling the free enzyme. Step (4): ADP is phosphorylated to form ATP. Reforming the oxidizing agent $NAD^+$ is the link to oxidation-reduction energy that I must outline here. My hypothesis claims that this process could have been realized in a sulfur-proteinoid system.

### Proteinoid Microspheres

The incomplete proteinoid construct of Section 2-1 can now be completed. Although some of this section is hypothetical, the experimental existence of

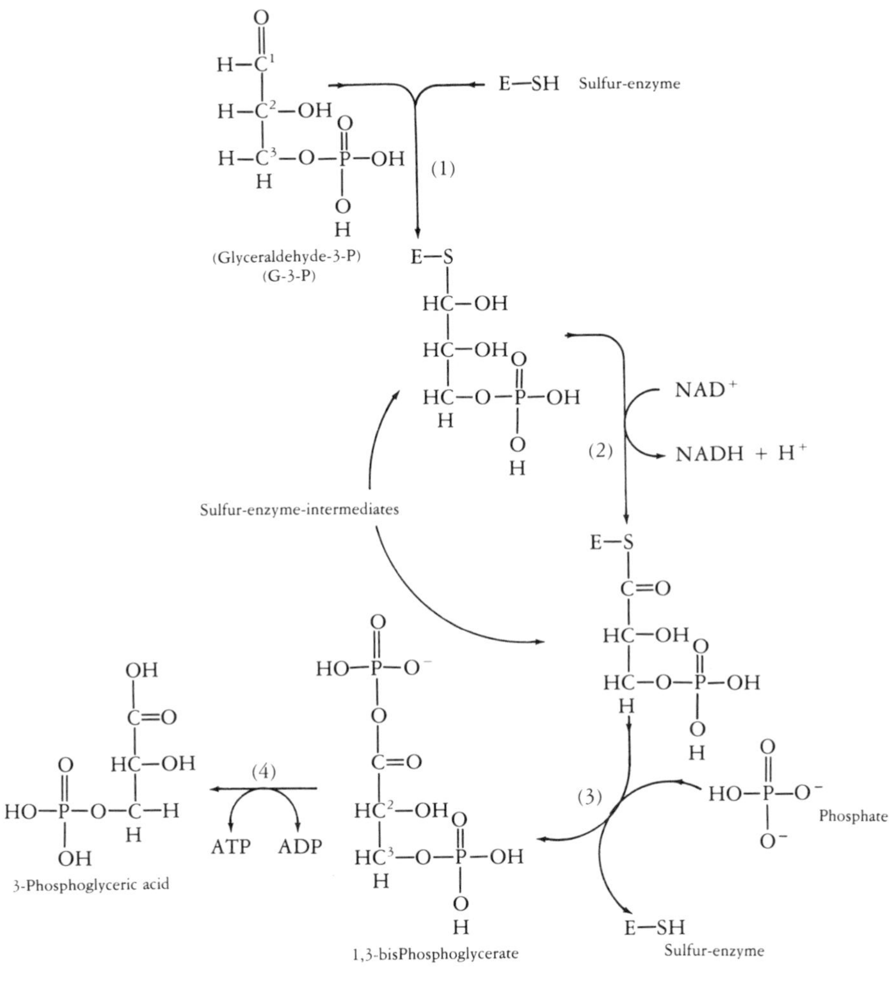

**FIGURE 2-6**
Glyceraldehyde-3-P → 1,3-bisphosphoglycerate + ATP, in contemporary organisms. Oxidative phosphorylation involves intermediation of a sulfur atom on a cysteine residue of the enzyme.

the proteinoids is not: Dry heating certain combinations of the 20 biological amino acids automatically results in amino acid chains called thermal proteins, or proteinoids. The lengths and compositions of these proteinoids depend on the composition of the amino acid mixture, the temperature, and other conditions of the experiment. The construct went only this far: splashing water solutions of amino acids onto hot igneous rock and then washing the resulting proteinoids back into the water.

***Proteinoids in the Laboratory*** Biochemists have discovered that when proteinoids dissolve in water, they spontaneously assemble (*self-assembly*) into spheres about one micrometer (μm) in diameter, called *microspheres* (Figure 2-7).

The microspheres enclose chains of amino acids and other solutes in the water in which they form. Although the proteinoids contain no lipids, the presence of hydrophobic amino acid residues causes the boundaries of the microspheres to act as membranes (some are even double layered). The boundaries tend to keep amino acid chains inside the microspheres; but sometimes other chemicals and protons can cross a boundary (though some of the more complex microspheres are even impervious to protons).

Proteinoids can act as hormones as has been shown experimentally by a proteinoid formed from six amino acids—glutamic acid, glycine, histidine, arginine, phenylalanine, and tryptophan. In a specific sequence, these amino acids comprise the active portion of the melanocyte-stimulating hormone; this hormone applied to a frog's skin causes black spots to appear. Black spots also

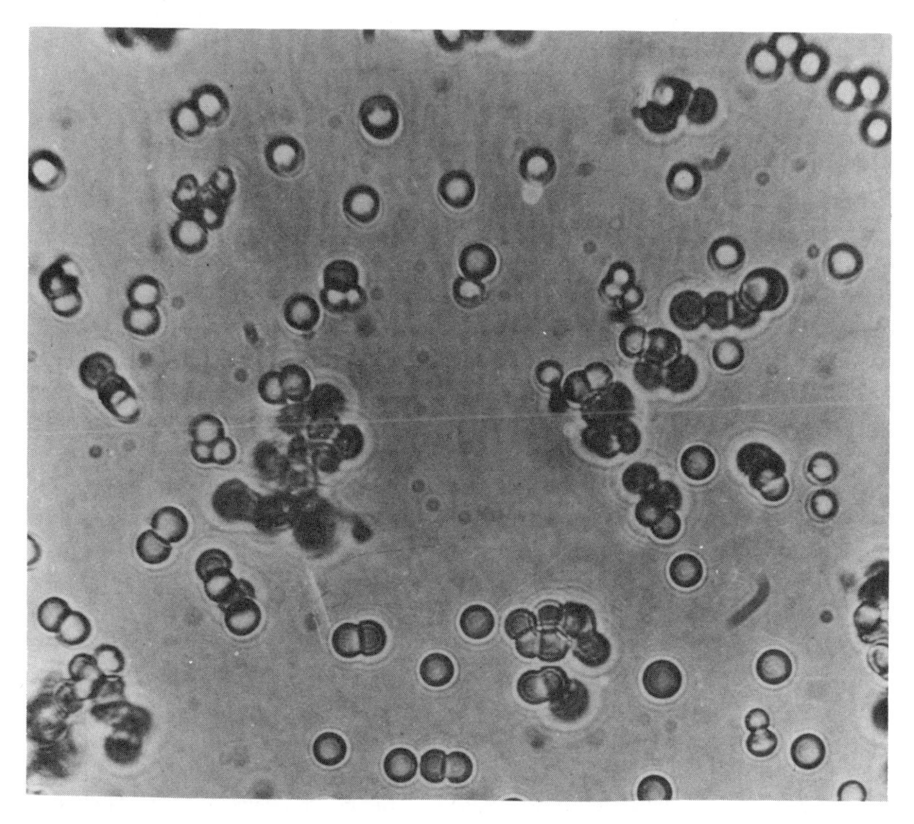

**FIGURE 2-7**
Microsphere photomicrograph. Note the remarkable uniformity in size (~2 micrometers diameter) of the protenoid microspheres. Each microsphere is self-assembled from approximately $10^{10}$ molecules of proteinoid. One gram of proteinoid yields $10^8$ to $10^9$ microspheres.

appeared when the experimenters applied the proteinoid to a frog's skin, because the proteinoid contained an abnormally large amount of the preferred specific sequence.

***Microspheres on the Prebiotic Earth*** On the prebiotic earth, proteinoid microspheres could have formed a metastable environment in which chemical processes could occur relatively protected from harsh geophysical processes. A proteinoid microsphere containing sulfur-proteinoids would provide an environment for the oxidation-to-phosphorylation process described earlier in this section. (If the sulfur atom is the key to the enzyme activity in this process, then the contemporary enzyme is simply a more efficient catalyst than the sulfur-proteinoid.) $NAD^+$ is a mixed oligomer (a dehydration condensate) that would have been available on a marginal basis in the primordial environment: It could have been present in some microspheres. If it acted cyclically—that is, if it was reoxidized—then small amounts would have been sufficient initially. The oxidation of NADH could have occurred by the action of one of the previously mentioned sources of oxidation-reduction energy; moreover, the membranelike activity of the proteinoid microsphere could have facilitated this coupling. A clue to how to accomplish this reoxidation comes from contemporary organisms.

The electron transport chains in contemporary cells, which are responsible for most of the energy harvested after glycolysis, use a quinone derivative to carry out a key step in the oxidation-reduction sequence (see Figure 2-5). The quinone simply moves between electron donor and electron acceptor. No protein appears to be required for its function. Figure 2-8 depicts the reoxidation of NADH, mediated by a quinonelike substance (Q), which is itself oxidized by the oxidation potential of ferric iron, Fe(III), which is assumed to have been abundant in the primitive environment and, therefore, near some microspheres. (The iron could be a residual of the iron catastrophe or could result from UV irradiation of organic iron compounds, which produces substances such as ferricyanide.) This mechanism also suggests that protons moved from inside to outside the microspheres, thereby generating a membranelike electrochemical potential. Protons could partially neutralize inorganic phosphate on the outside, thereby allowing it to penetrate the proteinoid boundary more easily (see Figure 2-9). The final step is the production of pyrophosphate, my hypothesized source of primordial phosphate bond energy.

The *Thiobacillus* mechanism for oxidizing $SO_3^{2-}$ to $SO_4^{2-}$ is perhaps even simpler. The mechanism used by this organism is, of course, catalyzed by enzymes (Figure 2-10); however, the acceptor of electrons from sulfite $SO_3^{2-}$ [step (1) in the figure], could have been iron(III) in the hypothetical scenario. In a primordial scheme based on this mechanism, I shall need only the formation of pyrophosphate $P \sim P$, instead of ATP, as in step (3). This requirement yields the simple model in Figure 2-11.

Once some mechanism, perhaps not unlike the one just proposed, coupled environmentally generated oxidation-reduction energy to the production of pyrophosphate in some proteinoid microspheres, the situation would have

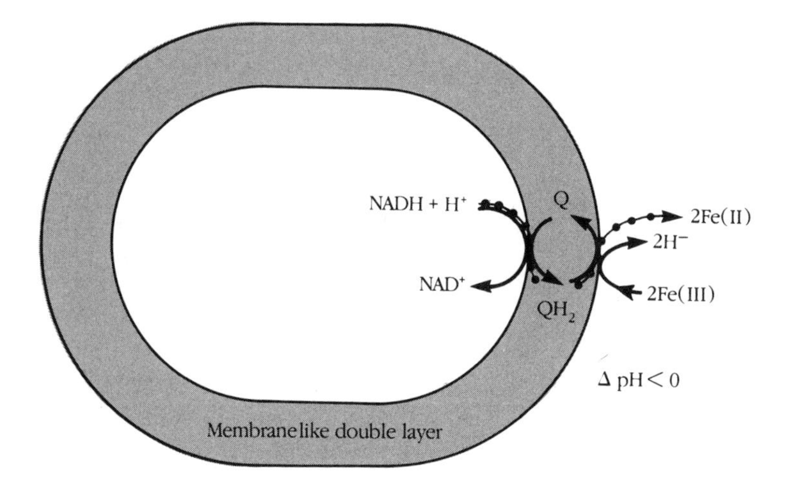


**FIGURE 2-8**
Reoxidation of NADH by quinone and Fe(III), in proteinoid microspheres. The quinonelike substance, denoted by Q, is one of the products of small molecule abiogenesis; it acts here as an electron and proton shuttle.

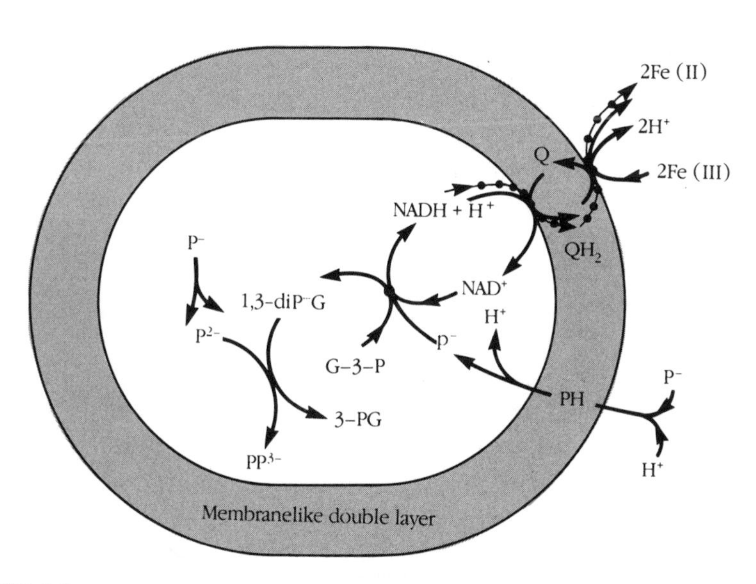


**FIGURE 2-9**
Integrated energy transduction, in proteinoid microspheres. G denotes glyceraldehyde (or glycerate); P, phosphate; and the large dot, the sulfur proteinoid catalyst for the phosphorylation of glyceraldehyde-3-phosphate. Dotted line shows electron movements.

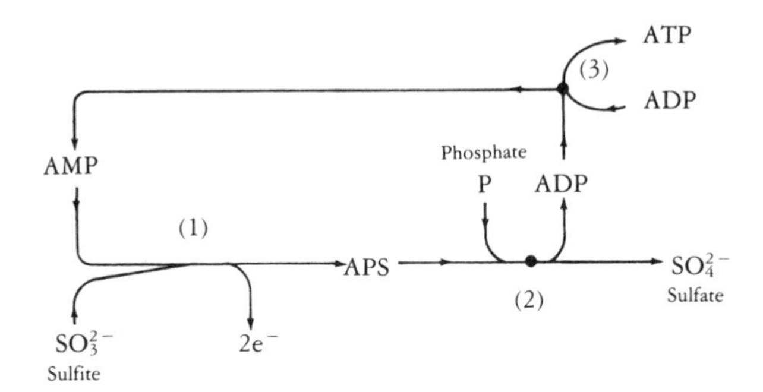


**FIGURE 2-10**
Sulfur oxidation in *Thiobacillus.* APS is adenosine-5′-phosphosulfate, also called active sulfate; dots denote enzymes.

been favorable for the activation and polymerization of monomers into oligomers and polymers. Perhaps one of the earliest consequences of abundant pyrophosphate would have been the generation of mixed oligomers that have the catalytic functions of coenzymes.

When considering energy metabolism, especially such sequences of reactions as those in Figure 2-3, in which five coenzymes participate, I have inferred that a stage of metabolism that was coenzyme dependent but virtually enzyme free (except for the proteinoids) evolved before anything like contemporary polymer synthesis arose. Although reaction rates in such systems would be much slower than in contemporary, enzyme-catalyzed organisms, they could easily be greater than the hydrolysis rates for small mixed oligomers (compare with the discussion of peptide hydrolysis in Section 2-1). With P ~ P energy driving this relatively simple system of reactions, a set of coenzymes could have evolved that would possess enough functional diversity to promote a connected system of reactions that also could regenerate the coenzymes. This would be uroboroslike behavior; but without genetics it could not be evolutionary.

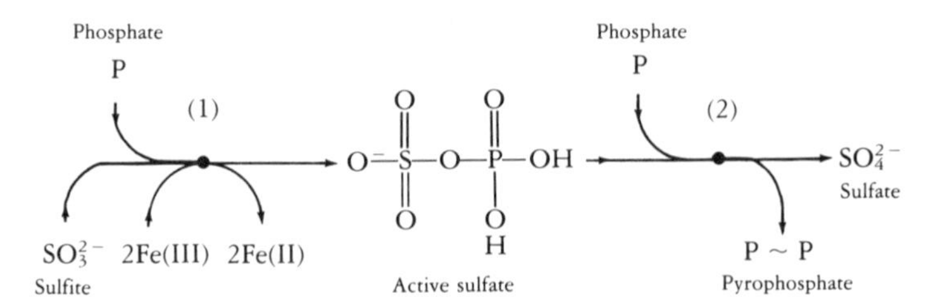


**FIGURE 2-11**
Coupled sulfite oxidation and formation of high-energy P ~ P, in proteinoid microspheres. Step (1), oxidative phosphorylation of sulfite to form active sulfate. Step (2), phosphorylation of phosphate to form P ~ P, with the release of sulfate. Standard oxidation-reduction potentials for these reactions at either acid or basic pH confirm that sufficient energy is available to generate P ~ P.

***Rates of Reactions*** The net result of energy-coupling mechanisms depends on the rates of the various competing processes. For example, thus far the model includes readily available pyrophosphate bond energy that either could degrade into heat through hydrolysis or could activate amino acids by linking with them to form energy-rich carboxyl phosphates. Which outcome dominates depends on rate, not equilibrium. Equilibrium favors hydrolysis of the pyrophosphates. Or, hydrolysis of the carboxyl phosphates could follow the successful activation of amino acids, which would produce the same net result as pyrophosphate hydrolysis. Rates are the determining factor. The hydrolysis of a dehydration bond in the absence of an enzyme may be as slow as $10^{-4}$ to $10^{-6}$ per second. Even if pyrophosphates hydrolyze as often as $10^{-2}$ per second, activation and polymerization could dominate (at $10^{-1}$ to $10^{3}$ per second), without approaching the rates typical of highly evolved modern enzymes ($10^{6}$ to $10^{10}$ per second). Activated amino acids could have polymerized before they hydrolyzed, even though hydrolysis was the ultimate equilibrium condition. Fundamentally, which rate dominates is an experimental question for which many important details, such as ion concentrations and pH are critical; however, the fact that life exists in its present form suggests that on the primitive earth phosphate energy drove the polymerization of amino acids rather than degrading into heat.*

## Primitive Genetics in the Microsphere

One gram of proteinoid can yield 100 million to a billion microspheres; their interaction with the prebiotic environment were myriads of natural experiments. I am proposing that in some of these experiments the production of phosphate bond energy was sufficient to polymerize amino acids, perhaps because the proteinoid involved was catalytic for peptide bond formation (Section 2-1). If some proteinoid could promote nucleotide polymerization with the 5′-3′ phosphodiester linkage, then small oligoribonucleotides (short chains of RNA) could have formed. Speculative detail about these steps is precluded by ignorance, but now the stage is set for the development of molecular genetics on a primordial level: Pyrophosphate-energized microspheres could have activated amino acids and produced short oligomers of RNA.*

In the next four subsections, I present speculations on how these microspheres could have developed a primitive genetics having the capability of evolving into a more complex system. Moreover, my scheme will suggest the ways in which many features of the contemporary mechanism could have their roots in this primordial one. I present these speculations partly because I feel they are plausible and interesting, but principally to demonstrate that the

* An ex post facto explanation of this circumstance has the same sort of impact that the 7.67-MeV second excited state of the $^{12}C$ nucleus had for stellar nucleosynthesis (see Section 1-1). No physical or chemical law is in question. The connection of universal laws with the particulars of the phosphate chemistry just described is remote.

* Cech's model (see Section 2-1) does not apply at this level because it requires a special, very large RNA as catalyst.

genetic apparatus, though complex, is not so complex that plausible origins for it cannot even be conceived. A current view exists that this problem of origin is insurmountable and that therefore life must have originated somewhere else where conditions were different (Crick, 1981). I reject such notions as defeatist and unimaginative.

***Primitive Polypeptide Formation*** Once again, the next step in formulating this model derives from the contemporary mechanism: In all known cases of gene-directed protein biosynthesis today, after ATP activates amino acids, tRNA synthetase immediately places them on their cognate tRNAs. The 2′—OH carboxyl-ribose ester linkage so formed is depicted in Figure 2-12. This ester bond is energy-rich to the same degree that the precursor amino acyl adenylate (the activated amino acid) was energy-rich. Thus the subsequent formation of a peptide bond between two activated amino acids is thermodynamically favored.

In this model of primordial gene-directed protein biosynthesis, I assume that *RNA served both as a primordial gene and as a primordial messenger RNA.* No DNA was involved, nor were ribosomes, tRNAs, or tRNA synthetases, which evolved later. First, pyrophosphate would activate the amino acids in a proteinoid microsphere to become amino acyl carboxyl phosphates. These would react with the 2′—OH groups of the RNA, forming amino acyl RNA esters (carboxyl-ribose esters), as in Figure 2-13, step (1). Unlike the tRNA case, in which these esters form only at the 3′ end, in this model they could form on any of the 2′—OH groups in the RNA. A modern RNA molecule usually has cations (positively charged ions), such as $M_g^{2+}$, bound to its phosphate groups, which are negatively charged; in this model, when an amino acid ester formed, the positively charged amino groups moved close to the RNA

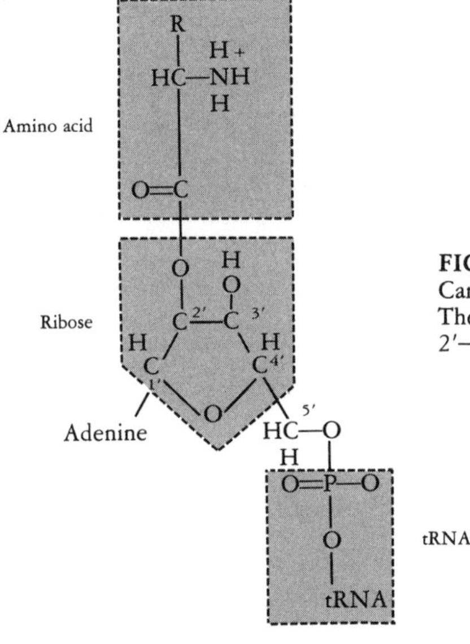


**FIGURE 2-12**
Carboxyl-ribose ester of tRNA, contemporary. The 3′ end of a t-RNA is shown, with the 2′—OH amino acyl ester at the top.

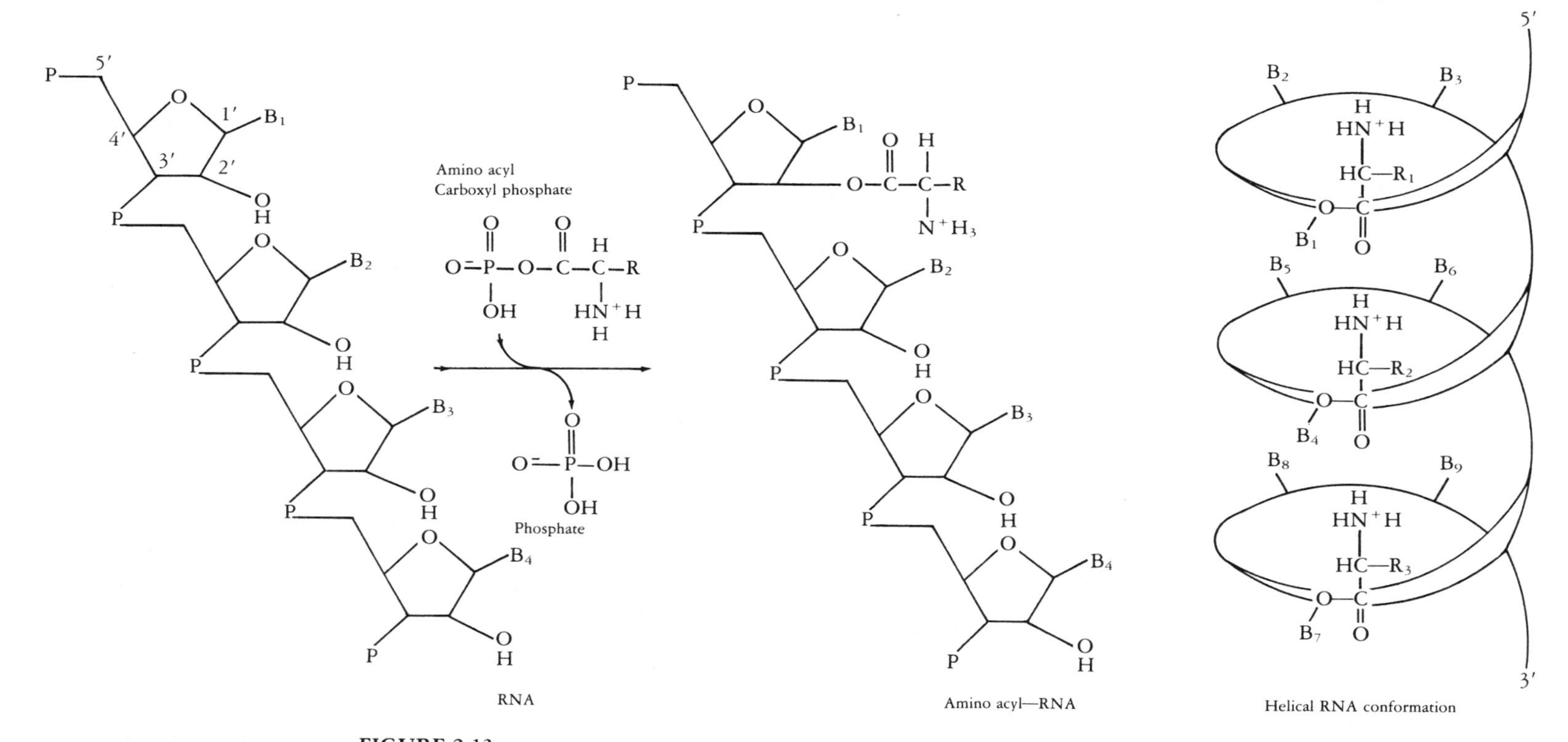


**FIGURE 2-13**
Amino acyl—RNA esters. Step (1): Activated amino acid monomer reacts with RNA. Step (2): Conformation change in RNA lines up amino acid-RNA esters for peptide bond formation.

phosphates, and displaced the metal cations. When enough esters formed, a conformation change occurred, as in step (2). The esterified RNA would form a right-handed helix of three nucleotides per turn, with the phosphates on the inside and the bases and amino acids on the outside. When amino acids esterified to every third ribose of RNA, they would be adjacent to each other in this helical conformation. Moreover, they would be adjacent in the sense that the amino group of one amino acid was adjacent to the ester carboxyl of the next amino acid located three riboses away in the 3′ direction. In the uncoiled conformation of RNA, amino acids that are three riboses apart are roughly 20 angstroms (Å) apart, and even amino acids on adjacent riboses are about 7 Å apart; these distances are too great for any chemical bonds to form. But in the model, the amino and carboxyl groups of some amino acids would be touching in the helical conformation and could then form the energetically favorable peptide bonds between them. This process would continue until all amino acids, three riboses apart, were bonded into a polypeptide, albeit a short one. To test the plausibility of these steps, I have built CPK and skeletal molecular models that accurately reflect the distances involved.

When the polypeptide was complete and left the RNA, the RNA would revert to its nonhelical conformation and could serve as a template for further protein synthesis—or for its own complementary copying (to be described below). The separation of the growing polypeptide from the RNA might have involved nothing more complex than spontaneous hydrolysis of the connecting carboxyl ester linkage.*

The base spacing in the model (like ribose spacing) would be three bases per amino acid, identical to the contemporary coding system, except that in the model it would be a consequence of the special helical conformation. The most plausible argument (Eigen, 1971) given for the three-base code in the contemporary mechanism is that two bases are not enough to code for 20 amino acids ($4 \times 4 = 16$ codons), and a two-base hydrogen bonding of tRNAs to mRNA is not strong enough to function smoothly or maintain specificity. On the other hand, a four-base code, although more than rich enough to account for the 20 amino acid alphabet ($4 \times 4 \times 4 \times 4 = 256$ codons), leads to too strong a bond between tRNA and mRNA and thus to too slow a mechanism. This leaves the three-base code as the best form.

But where is the coding in the primordial model? How could a specific amino acid bond (by esterification) to a specific base sequence on the RNA? I hypothesize that the esterification of an amino acid between two adjacent bases (for example, $B_1$ and $B_2$ in Figure 2-13) is selective, depending on the affinities of the two bases and the amino acid residue R. This would constitute a two-base code with three-base spacing (also reminiscent of the contemporary

* Several features of this model are of interest because of their similarity to the contemporary system. For example, the polarity of the protein produced in the model is such that the RNA is read from the 5′ end to the 3′ end while the protein is made from amino end to carboxyl end. This is identical to the polarity alignment in the contemporary mechanism. (If D-ribose is used, then L-amino acids fit better sterically than D-amino acids, which perhaps indicates the correlation between the two handednesses even if it says nothing about absolute handedness.)

code with its third-base redundancy, called wobble). A list of the first two bases of all codons of modern RNA codes for nearly all of the 20 amino acids used in biology appears in Table 2-4. When researchers are able to produce the right physicochemical conditions for this model in the laboratory, then this hypothesis of the origin of the code, particularly the specific code in Figure 2-13, can be tested experimentally.

The hypothesized coding mechanism offers some specificity of interaction between activated amino acids and the base sequences of RNA. This specificity need not possess the fidelity of the contemporary coding mechanism because failure among the myriad microspheres, which self-assemble so easily, is not costly in the biological sense. It need only favor one amino acid slightly more strongly than another. For example, lysine might bond between A and A only a few percent more often than histidine, and visa versa between C and A. This would produce a mechanism with a relatively high mutation rate (mutation is error) and would create the possibility among the billions of microspheres for rapid evolution.

*The Transition from No-Life to Life* Once microspheres existed that were capable of making RNA and then small proteins on this RNA, a particular

TABLE 2-4
*Two-Base Code*

| Doublet codon | | | Amino acid |
|---|---|---|---|
| 5′ | AA | 3′ | Lysine |
| | AC | | Threonine |
| | AG | | Serine or arginine |
| | AU | | Isoleucine |
| | CA | | Histidine |
| | CC | | Proline |
| | CG | | Arginine |
| | CU | | Leucine |
| | GA | | Aspartate or glutamate |
| | GC | | Alanine |
| | GG | | Glycine |
| | GU | | Valine |
| | UA | | End ("punctuation") |
| | UC | | Serine |
| | UG | | Cysteine |
| | UU | | Phenylalanine |

RNA sequence could, by chance, code for a short protein that would catalyze RNA copying (that is, would cause nucleotide polymerization on an existing RNA molecule). The second copy of this RNA sequence would reproduce the original RNA. In this way, the model RNA could also act as a reproducible gene. This development could result in a system that is self-generating—the initiation of uroboros. This model thus provides a plausible scheme for the transition from no-life to life. Figure 2-14 extends the ideas in Figures 1-5 and 2-1 to this scheme.

## Evolution to Greater Complexity

How could such a system have evolved into the contemporary complexity of ribosomes, tRNAs, and synthetases? This question is difficult, and no experimental evidence exists to aid or discipline speculation. (Nevertheless, researchers have proposed models for the origin of tRNAs and even tRNA synthetases; see Kuhn, 1983). The evolutionary capacity of this primitive model is the essential ingredient: Imperfect gene replication or translation would have conferred mutability, which would have led to the selection of systems that coupled and used energy more efficaciously. Evolution poses a

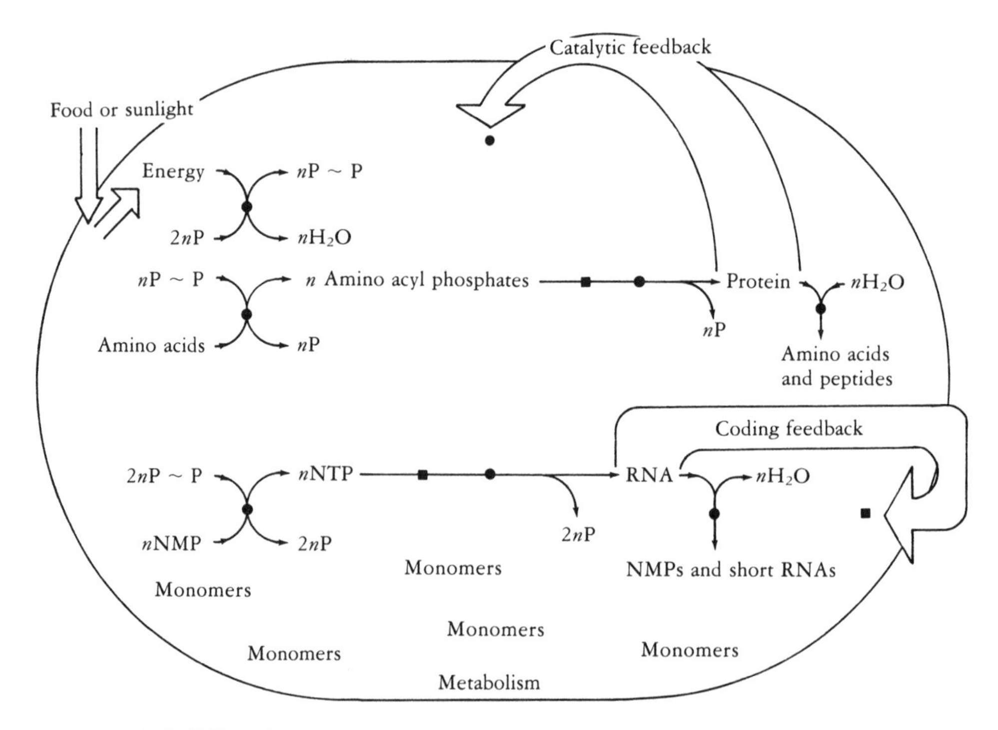


**FIGURE 2-14**
Feedback structure for polymer synthesis, in proteinoid microsphere. Dots denote protein catalysts; squares denote RNA "catalysts" (that is, primitive genes and messengers, simultaneously).

problem that is clearly related to the initiation of the uroboros. If new capabilities are to evolve, they must accumulate one at a time, not in large numbers. For example, the 20 tRNA syntheses required by a contemporary system could not have been evolved all at once.

A plausible hypothesis regarding the emergent properties of tRNA synthetases exhibits this idea of one-by-one additions. It also provides an evolutionary explanation for the observed diversity of mechanism and structure in contemporary synthetases.

***Synthetase Evolution, One at a Time*** Assume that the primordial uroboros has begun to function and therefore that energy not only is abundantly available but is being converted into pyrophosphate at a high rate and is being used for activation of amino acids and their polymerization at a rate that is faster than that of hydrolysis. One product of this energy-abundant state would be the production of RNAs, some of which would serve as genes and messengers for crucial protein synthesis. The rest of the RNA would accumulate and hydrolyze into shorter fragments, including three-nucleotide fragments. Longer fragments might find a stable conformation in a hairpin shape. These would become the primitive tRNAs.

Now suppose that one of the various RNAs is a gene for a triplet-binding protein, a *single* protein that can bind a sequence of any three nucleotides. For example, the locations of two arginine residues could be such that they bind the phosphates of the nucleotide triplets by electrostatic bonds. Note that this binding would not be specific to the triplet sequence but only to the triplet length (Figure 2-15). As a consequence of the formation of just one such protein, a set of triplet-binding protein-nucleotide complexes could arise,

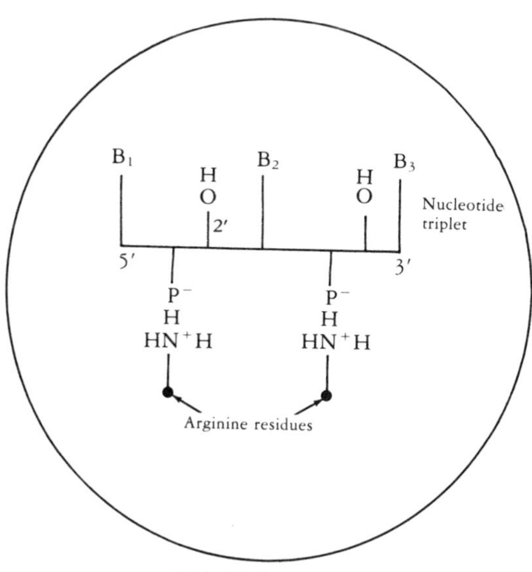


**FIGURE 2-15**
Primitive synthetase, in proteinoid microspheres. The postulated arginine residues might be lysines instead. These residues bind the negatively charged phosphates of nucleotide triplets.

each composed of the same protein part but with differing nucleotide triplets. These protein-nucleotide complexes will serve in this model as primitive tRNA synthetases.

A primitive tRNA synthetase would have to recognize and bind to both an activated amino acid and a primitive tRNA that contained the complementary three bases (cognate tRNA). The primitive tRNAs were the hairpin shape fragments mentioned above. At the loop end of the hairpin, three bases would serve as the anticodon, which would be recognized by and bound to the primitive synthetase by base-pairing to its bound triplet (Figure 2-16). Amino acid recognition is another postulated function of the bound nucleotide triplet of the primitive synthetase. Recognition would take place by the preferential bonding (esterification) of the carboxyl group of the activated amino acid to the 2′—OH of the ribose sugar of the synthetase triplet, between the two bases at its 5′ end. This procedure is identical to that postulated earlier for the direct reading of an RNA messenger.

Next, the tRNA synthetase would attach the amino acid to the 3′ end of the primitive tRNA. I postulate that the primitive tRNA would change conformation so that its 3′ end would contact and bond to the amino acid ester (transesterification). Then this amino acylated primitive tRNA would leave the synthetase with the cognate amino acid attached. Such charged tRNAs could then read the RNAs that were previously read directly, provided that primitive ribosome (perhaps a portion of the microsphere's inner membrane) has evolved to coordinate this reading. The advantage of an indirect mechanism

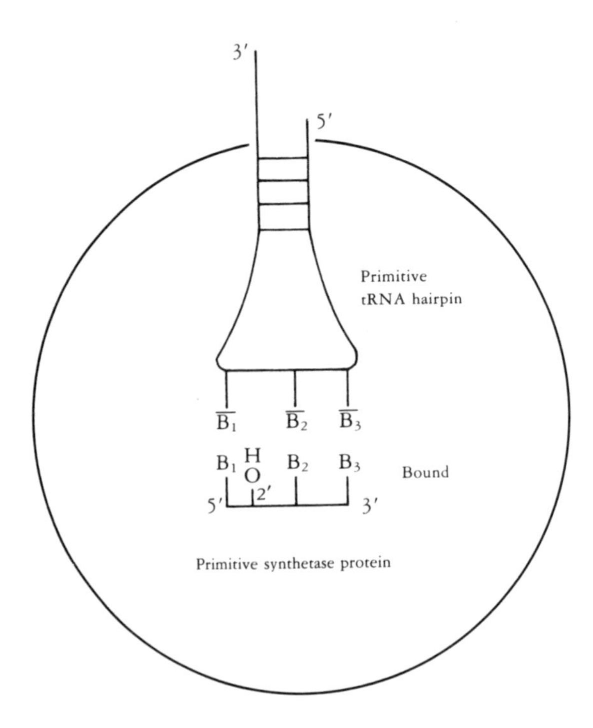


**FIGURE 2-16**
Primitive tRNA, in proteinoid microspheres. The 2′—OH group between bases $B_1$ and $B_2$ is the locus for amino acid esterification. The cognate tRNA hairpin loop is in the binding position. Once an amino acid esterifies to the 2′—OH group, the 3′ end of the hairpin can bend down for transesterification to the tRNA.

of translation of the RNA over its direct reading would be rate and fidelity. These and many other details of these mechanisms require explication, but the most important point is that it *all would derive from a single protein.* Subsequently, other proteins could arise (one at a time) that would perform the recognition and binding functions better than the simple, triplet nucleotide-protein complexes; new proteins would improve on old ones and finally replace the primitive synthetases (again, one at a time, and often with redundant or overlapping functions). Ultimately, the new proteins would function without bound nucleotide triplets, and the system would then have pure protein synthetases, just as contemporary organisms do.

This hypothetical scheme would lead to the accumulation of a large number of proteins having synthetase function. As they accumulated, the mechanism of translation would become more complex, and yet the code itself would remain fundamentally unchanged. With greater fidelity, the third base of the code could become useful in coding for previously unused amino acids, such as tryptophan and methionine; but for the most part the primitive two-base code would be preserved.

## Perspective on Uroboros Initiation

The environments provided by the proteinoid microspheres were far more felicitous for the transition from no-life to life than was that provided by thermal equilibrium. Geophysical energy sources enriched the earth's environment enormously. The difference between the dynamic states of the energy-driven microspheres and thermal equilibrium for the primordial dozen elements was the significant difference. In the context of microsphere—with their proteinoid chains, membranelike boundaries, and dissolved phosphates and other chemicals—the emergence of short RNA molecules does not seem so miraculous as it does in the context of events in an unprotected geophysical setting. This is the sense of Lipmann's answer to Delbruck's quest (see Introduction) for the transition from no-life to life: The coupling of oxidation-reduction energy to phosphate bond energy "might have been the first event on the way to life."

In some sense, energy-driven microspheres seem already alive, their behavior and great number imparting the aspect of a living state to the earth's lithosphere, a viewpoint that invokes the flavor of James Lovelock's Gaia hypothesis. Instead, this book views this first lifelike form as a stage of organic development, in which existed a primitive molecular genetics based on the RNA of the primitive uroboros.

This definition of life specifies a mechanism that depends on phosphate bond energy. Any experimental paradigm designed to study the origin of such mechanisms must also study ongoing, energy-driven polymerizations. The effects on the polymerization mechanism itself of the emergent properties of the polymers produced will be the focus of such studies. The bioengineering technology needed for such studies is already available.

## 2-4. THERMODYNAMIC ANALYSIS OF THE UROBOROS

A flux of energy was the impetus behind the initiation and maintenance of the dynamic state of matter that we recognize as living. However, evolution implies that life is more than a merely self-begetting dynamic state; to be recognized as living, a self-begetting state must slowly alter its character through accumulation of many small changes. Thus it cannot be a perfectly self-replicating uroboros. After all, a perfect uroboros at the earliest heterotroph level could never have become autotrophic, which is what primitive living matter has done.

### Equilibrium: Thermally Isolated versus Thermally Buffered System

A fundamental dichotomy exists between an equilibrium state of the primordial elements and the state of an energy-driven living system. The equilibrium state of the primordial elements occurs at the temperature of the crustal environment in which they exist. Some small differences in temperature exist between different local environments, and small temperature variations occur with time in any location; but basically the equilibrium state is characterized as thermal equilibrium, and the planet's crust acts as the thermal reservoir, with water serving as a thermal buffer. The equilibrium of a totally isolated system, not in contact with a thermal reservoir, is different. These two cases require some discussion about the distinction between *free energy* and *entropy*.

The *second law of thermodynamics* governs the equilibrium state of matter. For an *isolated system*, this law says that the equilibrium state is the one with the *maximum entropy for a fixed total energy*; for a system in *contact with a thermal reservoir* the second law says that the equilibrium state is the one with *minimum free energy*. These two statements are not in conflict; they are manifestations of the same law in two different situations.

From a statistical mechanical point of view, entropy is a measure of a system's order or disorder; increased entropy corresponds to decreases of order, or increases of disorder. Energy flowing into an isolated system drives the system away from its equilibrium state; and since its equilibrium state in this case has maximum entropy, *the effect of the energy flow is to decrease the entropy,* that is, to increase the system's order. This phenomenon is called *energy-flow ordering* or, sometimes, "creating order out of chaos." These ideas are valid for a system that is isolated except for the energy input.

But the setting for the emergence of life was not an isolated system; it was a thermally buffered. Energy flowing into a thermally buffered system drives the system away from its equilibrium state, which means that its *free energy increases.* Free energy has two components: the so-called internal energy and the entropy. Thus a free energy increase may be due to a change in internal energy, not in entropy, making it wrong to think in terms of order out of chaos when dealing with a thermally buffered system.

## Implications of Thermodynamic Dichotomy for Biopolymers

In constructing the primitive uroboros, as described in Section 2-3, I have used two types of polymers: proteins, which are catalysts and structural components, and polynucleotides, which are components that control and direct the synthesis of proteins. In protein synthesis, the genetic apparatus provides the mechanism by which the sequence of amino acids in a protein is determined, residue by residue. To conclude that the sequences of amino acids in a chain would be random if it were not for the careful control imposed by genes is natural, seductively so; but it leads to the notion of order out of chaos (lower entropy from higher entropy), which is invalid for energy-driven thermally buffered systems. In such systems an increase in free energy occurs. For peptide formation, the change is mostly an increase in internal energy rather than a decrease in entropy. (Indeed, dipeptide formation sometimes shows an increase in entropy (over free monomers) when quantitative calorimetry is used.) Because polynucleotide formation results largely from increases in internal energy, driving the transition from monomers to polymers requires activation free energy. Phosphate bond energy has its source in this increase of free energy and internal energy, not in a decrease of entropy.

The thermodynamic issue of the amino acid sequence in proteins is not purely an issue of entropy. The amino acids in a protein cannot be thought of as colored beads on a string, nor can the order of the sequence be assessed by calculating the number of arrangements of colored beads that are mathematically possible. Experiments on the thermal polymerization of amino acids (which makes the proteinoids I used in constructing the primitive uroboros) provide the essential clue. Proteinoids are not random sequences of amino acids: They show a pronounced tendency to form specific sequences, presumably for chemical reasons since the various amino acids have different residues that have different chemistries. Thus *polymerization is a chemical ordering, not a mathematical one.* To think otherwise is to ignore a vast quantity of experimental data.

One example of this intrinsic tendency to form specific sequences is the experiment described in Section 2-3, in which six amino acids, in a specific sequence, comprise the active sequence of melanocyte-stimulating hormone (Fox and Wang, 1968). The chance of obtaining the active sequence when these six amino acids are thermally polymerized would be only 1 in nearly 50,000 if the process were purely random. Instead, the yield of this product is 20 percent (10,000 in 50,000), or ten thousand times the amount predicted by statistics. The thermally produced amino acid sequence has a similar activity to the hormone fragment. Not only is this sequence dominant in a non-gene-directed chemical reaction, but it is even a biologically active sequence.

In my uroboros model, the proteinoids possessed catalytic functions such as energy coupling, activation of monomers, and polymerization. If the amino acid sequences produced were mathematically random, and if the catalytic functions corresponded to specific sequences, then the odds would be very much against the formation of functional proteinoids. If, instead, some sequences are chemically much more likely to occur than others, and if these are the catalytic ones, then the model primordial uroboros has a chance.

### Evolutionary Capability

As evolution proceeded, slowly establishing an increasingly faithful genetic apparatus, the primordial uroboros would develop the ability to produce amino acid sequences that performed catalytic functions, which were needed for developing and refining metabolism and polymerization. The genetic apparatus seems to have made possible the *diversification of sequences* for diverse functions; it has not selected a limited number of sequences out of an astronomical number of combinatoric possibilities.

What sets evolution, as a process, apart from other energy-driven processes is the very special properties of polymers, not just the energy input. The feedback effects of catalysis and regulation by proteins and of genetic memory by polynucleotides enable the energy flows to evolve complexity. A system without these emergent properties of biopolymers, even if it is energy-driven, will not have evolutionary capacity.

## REFERENCES


**Section 2-1**

Ballard, R. D., *Exploring our Living Planet,* The National Geographic Society, Washington, D.C., 1983.

Cech, T., "A Model for the RNA-Catalyzed Replication of RNA," *Proc. Nat. Acad. Sci. U.S.* 83 (1986): 4360.

Fox, S. W., and K. Dose, *Molecular Evolution and the Origin of Life,* Marcel Dekker, New York, 1977.

Lipmann, F., "Projecting Backward from the Present Stage of Evolution of Biosynthesis," in *The Origins of Prebiological Systems and Their Molecular Matrices,* edited by S. W. Fox, Academic Press, New York, 1965.

Lipmann, F., *Wanderings of a Biochemist,* Wiley-Interscience, New York, 1971.

Horowitz, N. H., "On the Evolution of Biochemical Synthesis," *Proc. Nat. Acad. Sci. U.S.* 31 (1945): 153.

Miller, S. L., and L. E. Orgel, *The Origins of Life on the Earth,* Prentice-Hall, Englewood Cliffs, N.J., 1974.

**Section 2-2**

Fox, R. F., *Biological Energy Transduction: The Uroboros,* John Wiley, New York, 1982. The material in this section is a partial condensation of material treated in part II of this earlier book, which contains additional information.

Lewin, B., *Genes,* John Wiley, New York, 1986.

Stryer, L., *Biochemistry,* 2nd ed., W. H. Freeman, New York, 1981.

Watson, J. D., *Molecular Biology of the Gene,* 3rd Edition, Benjamin, New York, 1975.

### Section 2-3

Ballard, R. D., *Exploring our Living Planet,* The National Geographic Society, Washington, D.C., 1983.
Crick, F., *Life Itself,* Simon and Shuster, New York, 1981.
Eigen, M., "Self-organization and the Evolution of Biological Macromolecules," Naturwissenschaften 58 (1971): 465.
Fox, R. F., *Biological Energy Transduction: The Uroboros,* John Wiley, New York, 1982. See comment to reference in Section 2-2. This time Part III is relevant.
Kuhn, H., and J. Waser, "Self-organization of Matter and the Early Evolution of Life," in *Biophysics,* edited by Hoppe, Lohmann, Markl, and Ziegler, Springer-Verlag, Berlin, 1983.
Kulaev, I. S., and V. M. Vagabov, "Polyphosphate Metabolism in Micro-Organisms," in *Advances in Microbial Physiology,* Vol. 24, Academic Press, London, 1983.

### Section 2-4

Fox, S. W., and C.-T. Wang, *Science* 160 (1968): 547.
Prigogine, I., *Order out of Chaos,* Bantam Books, New York, 1984.

CHAPTER

# 3

# Self-Assembly and Control

During the billion or so years after the first uroboros emerged 3.8 billion years ago, metabolism developed, along with the accumulation of genes, now part of a sophisticated genetic apparatus for protein biosynthesis with contemporary complexity. Catalytic ability diversified with increasing numbers of distinct genes, and eventually modern photosynthesis arose. This chapter does not chronicle this development in detail but instead focuses on a few emergent properties of proteins and polynucleotides, specifically the regulative mechanisms for protein synthesis and function.

In Sections 3-1 and 3-2, an account of how self-assembly creates complexes of molecules with emergent properties stresses the effect of water and the thermodynamics of self-assembly; both could have been operative in the primitive uroboros. Section 3-3 surveys contemporary examples of function regulation in self-assembled macromolecular complexes and demonstrates the prevalent occurrence of phosphorylation and dephosphorylation as regulatory mechanisms; it concludes with speculation about the role of phosphate in the development of primitive control mechanisms. Section 3-4 discovers a new role for phosphate as an energy storage molecule for rapid processes, a role that could have made major evolutionary developments possible in multicellular organisms.

## 3-1. SELF-ASSEMBLY

Water is a major molecular component of cells, serving as the solvent for essentially all cellular chemical reactions. It also serves as a thermal buffer and, to a lesser extent, as a pH buffer.

### Examples of Self-Assembly

After their synthesis, many polymers self-assemble into aggregate structures such as enzyme complexes and membranes. The enzyme complexes have multiple subunits, typically two or four, but some have prodigious numbers of subunits. The complex that catalyzes the conversion of pyruvate into acetyl—CoA in *E. coli* (see Figure 2-3) contains a total of 72 subunits of three different types and has a molecular weight of 4,440,000. After experimenters completely disassemble this complex in the laboratory, it can self-assemble to full activity.

The organized arrays of proteins in the electron transport chains in bacteria (see Figures 2-5) also self-assemble. The enzymes of the tricarboxylic acid cycle in eukaryotes self-assemble within the membranes of mitochondria. Even the complicated ribosome self-assembles.

Membranes self-assemble from lipid and protein precursors. Lipid (usually phospholipid) vesicles can self-assemble from lipid-water mixtures in the laboratory, where it is now feasible to self-assemble protein complexes that can function in self-assembled lipid vesicles. For example, ATPase complexes that have self-assembled in artificial lipid vesicles can make ATP from ADP and P when appropriately energized. Researchers have also functionally reconstituted transport mechanisms (such as for ions and simple molecules) in vesicle systems. Most of these assemblies require no added free energy and usually need no special catalytic agents.

## Weakly Bonded Complexes

The molecular structures discussed in Chapter 2 were held together primarily by strong covalent bonds (thermally stable at temperatures around 300 K) that provide the "backbone" structure of macromolecules. Weak, thermally labile bonds, such as hydrogen bonds, hydrophobic bonds, and van der Waals bonds provide specificity of interaction and conformational stability. The same weak bonds that endow proteins and polynucleotides with their ability to make specific interactions also stabilize aggregations or complexes of macromolecules. Weakly bonded assemblies of proteins, and of proteins with polynucleotides, abound in cells. The cell itself is such an assembly, composed of many subassemblies and surrounded by a self-assembled membrane.

Because self-assembly uses weak bonds, it possesses remarkable properties. Interactions can be specific and sufficiently stable, but not so stable that thermal perturbations cannot break the subunits apart. If the energy of a bond is $\Delta E$ in kilocalories per mole, then the expectation time for bond breaking (the time in which rougly half the complexes will break apart) is given by Boltzmann's formula (Schrödinger, 1944),

$$t = \tau \exp\left(\frac{\Delta E}{RT}\right)$$

in which $\tau$ is the reciprocal of the collision frequency, around $10^{-13}$ s for the solvent water at $T = 300$ K. For a weak bond with $\Delta E \sim 10RT_{300}$, the expectation time for bond breaking is $t \sim 2 \times 10^{-9}$ s. However, for three simultaneous weak bonds of equal strength, this exponential formula yields an expectation time for breaking all three bonds of $t \sim 1$ s, a long time on the biochemical time scale. This explains the stability of conformations with multiple weak bonds.

To make three simultaneous weak bonds, complementary groups on the surfaces of polymer subunits must be located in close proximity so that bonding takes place spontaneously. This obligatory complementarity confers specificity. Many nonspecific interactions do occur between subunits, but they involve only one weak bond and are neither stable nor specific.

## The Effects of Water

Water molecules take part in all biological self-assembly processes. The dispersed subunits in a cell, whether proteins or polynucleotides, are surrounded by organized water structures because water hydrogen-bonds readily with many chemical groups as well as with other water molecules. The R groups (residues) on a subunit that do not have specific weak bonds to complementary residues on another subunit will bond to water molecules, which will be displaced when further specific weak bonding between subunits takes place. Consequently, free energies of formation for self-assembly from dispersed subunits depend on the effects of water molecules at the molecular level.

Recall that even though subunit polymers are subject to hydrolysis, the rate of hydrolysis is slow (in the absence of enzymes); therefore, in their interactions with water molecules in the context of self-assembly they can be thought of as metastable subunits. After self-assembly, many subunits are even less likely to hydrolyze because much of their surface is sequestered in complexes away from water molecules.

Both the self-assembled structures and their functions are emergent properties; that is, they are unpredictable from the nature of energy flow through the primordial dozen elements.

### Free Energy and Entropy Considerations for Self-Assembly

Let the initial state for self-assembly be dispersed subunits, and let the final state be the aggregated complex. The free energy change $\Delta G$ is defined by

$$\Delta G = G_{\text{final}} - G_{\text{initial}}$$

that is, $\Delta G_{\text{self-assembly}} = G_{\text{aggregated}} - G_{\text{dispersed}}$. Thermodynamics tells us that this free energy change (Gibbs free energy) is related to changes in internal energy $\Delta U$, entropy $\Delta S$, and volume $\Delta V$ by the identity

$$\Delta G = \Delta U + P\,\Delta V - T\,\Delta S$$

in which $P$ is the pressure and $T$ the absolute temperature. Take $P$ to be atmospheric pressure and ignore the $P\,\Delta V$ term since $\Delta V$ here is relatively small.

The second law of thermodynamics (see Section 2-4) for an isolated system states that for a spontaneous process $\Delta S > 0$, whereas for a thermally buffered (and constant $P$) system the second law states that $\Delta G < 0$. Biological systems are thermally buffered. Experimentally, $\Delta G < 0$ is always true for the self-assembly of a complex from its dispersed subunits. Nevertheless, observers often feel that self-assembly must be paradoxical because the transition from dispersed subunits (initial state) to aggregation (final state) appears to be in the direction of greater organization, or increased order. As noted at the end of Chapter 2, increased order means an apparent decrease in entropy (that is, $\Delta S < 0$); but this is not necessarily in conflict with the second law because biological systems are not isolated. Moreover, $\Delta G < 0$ and $\Delta S < 0$ are compatible provided that

$$|\Delta U + P\,\Delta V| > T\,\Delta S$$

and

$$\Delta U + P\,\Delta V < 0$$

But the rationalization just advanced is not necessary. The appearance of ordering of subunits during self-assembly is real, but the disordering of water molecules is greater. The use of notation such as $\Delta G$ and $\Delta S$ is misleading if investigators say that $\Delta G < 0$ for the assembly process and $\Delta S < 0$ for the subunits. In reality, the $\Delta G < 0$ statement implicitly includes contributions from water molecules, whereas the $\Delta S < 0$ statement explicitly excludes such contributions. Investigators should include water contributions explicitly in both statements. To do so requires enhancing the notation so that $\Delta G = \Delta G_s + \Delta G_w$ and $\Delta S = \Delta S_s + \Delta S_w$, where the subscripts s and w denote subunits and water, respectively.

Careful analysis of the assembly process shows that $\Delta S_s < 0$ is true, primarily due to the loss of translational entropy of the subunits in going from the dispersed state into the aggregated state. In other words, dispersed subunits are free to move in space, independently of each other, whereas aggregated subunits move, or translate, together. This shows up as a decrease in translational entropy, the size of which dwarfs entropy contributions from rotational motion, vibrational motion, and so on. Because visual impressions are dominated by translational motion, the observer considers the aggregated state to be more ordered and the entropy to be decreased. The observer, however, cannot see the water molecules (unless very special technical tricks are used) and so does not consider $\Delta S_w$.

The internal energy term $\Delta U$ also must be separated into $\Delta U_s$ and $\Delta U_w$. These terms account for the energy in weak bonds. Looking only at $\Delta U_s$ yields $\Delta U_s < 0$ because weak attractive bonds bind the subunits together. However, the complementary residues in the complex were initially bonded to water by weak attractive forces that aggregation eliminated. Thus the water contribution is $\Delta U_w > 0$. Moreover, since both the water-to-subunit bonds and the complementary subunit-to-subunit bonds are of comparable strength, $\Delta U = \Delta U_s + \Delta U_w \sim 0$. (As stated above, $P \Delta V$ is ignorable.) Therefore, $\Delta U + P \Delta V < 0$ is not the case; instead, $\Delta U + P \Delta V \sim 0$.

So far, $\Delta U + P \Delta V \sim 0$ and $\Delta S_s < 0$; so $\Delta G = \Delta U + P \Delta V - T \Delta S$ looks as though it will be greater than zero ($-\Delta S_s > 0$ if $\Delta S_s < 0$). But this equation does not yet include $\Delta S_w$ in $\Delta S$, which is the key term because during aggregation the displaced water molecules gain translational entropy, making $\Delta S_w > 0$. Because several water molecules are usually bound to every subunit, $| \Delta S_w | > | \Delta S_s |$. The overall effect is that $\Delta S > 0$; so $\Delta G < 0$ because of the increased translational entropy of displaced water molecules. Thus the assembly process is "entropy driven," that is, driven by the entropy increase of water.

## 3-2. MOLECULAR MORPHOGENESIS AND CONTROL MECHANISMS

Self-assembly does not require added energy. Once the subunits are synthesized, their thermal motions bring them together, whereupon they spontaneously link by means of multiple specific weak bonds. The structures they form and the unexpected properties of those structures are examples of emergent properties, which are observed to attend all transitions in a structural hierarchy. A key emergent property is the capacity for control, on regulation. This section presents four different regulatory mechanisms: allostery (ATP and ADP case), phosphorylation-dephosphorylation (glycogen case), hormone and cyclic AMP cascade (glycogen case), and end-product inhibition (amino acid case). Biochemistry textbooks provide many others. Some readers may prefer to skip these examples and move to the end of the next section, which discusses primitive mechanisms and general principles.

### Regulation of ATP and ADP

The catalytic activity of many enzymes in metabolic pathways is regulated. For example, when a cell has plenty of ATP, a feedback control mechanism inhibits the ATP-generating pathways. The use of ATP leads to a buildup of ADP, which acts as an activator of the same pathways (Figure 3-1). The affected enzyme usually contains two sites, one that is specific for the substrates of the pathway it catalyzes and one that is specific for the effector (regulatory agent), in this case ATP or ADP. In many enzymes, these two sites are located on separate subunits of the complex: The effector binds to one subunit, causing a conformational change that is communicated to the catalytic subunit through their mutual weak bonding. This is called *allosteric* (other shape) regulation (Figure 3-2).

### Glycogen Metabolism

A remarkably prevalent way of causing conformational changes among subunits is to add and remove phosphate groups. Since the initiation, development, and evolution of life are based on phosphate bond energy, it is not surprising to find that control mechanisms are dominated by phosphate transfers. The metabolism of glycogen, a polysaccharide used by many types of animal cells for the storage of energy-yielding carbohydrate, is a sophisticated example of cellular energy storage and use and of regulation of such processes by phosphates.

***Glycogen Synthesis*** Glycogen synthesis is a response to excess glucose after unused ATP inhibits glycolysis (Figure 3-3). The polymerization of glucose into glycogen requires the activation of glucose monomers by uridine-5′-triphosphate (UTP) to form UDP—glucose, which adds them to the growing polymer. [UTP is generated from its precursors (UMP and UDP) by phosphorylations that are driven by ATP. In protein synthesis a parallel role is played by guanine-5′-triphosphate (GTP), which is also generated from ATP. In phospholipid synthesis an identical role is served by cytidine-5′-triphosphate (CTP). All of these nucleotide-triphosphates are required for RNA synthesis. In every case, the ultimate source of energy is ATP.] This activation of glucose monomers by UTP is analogous to the activation of amino acids

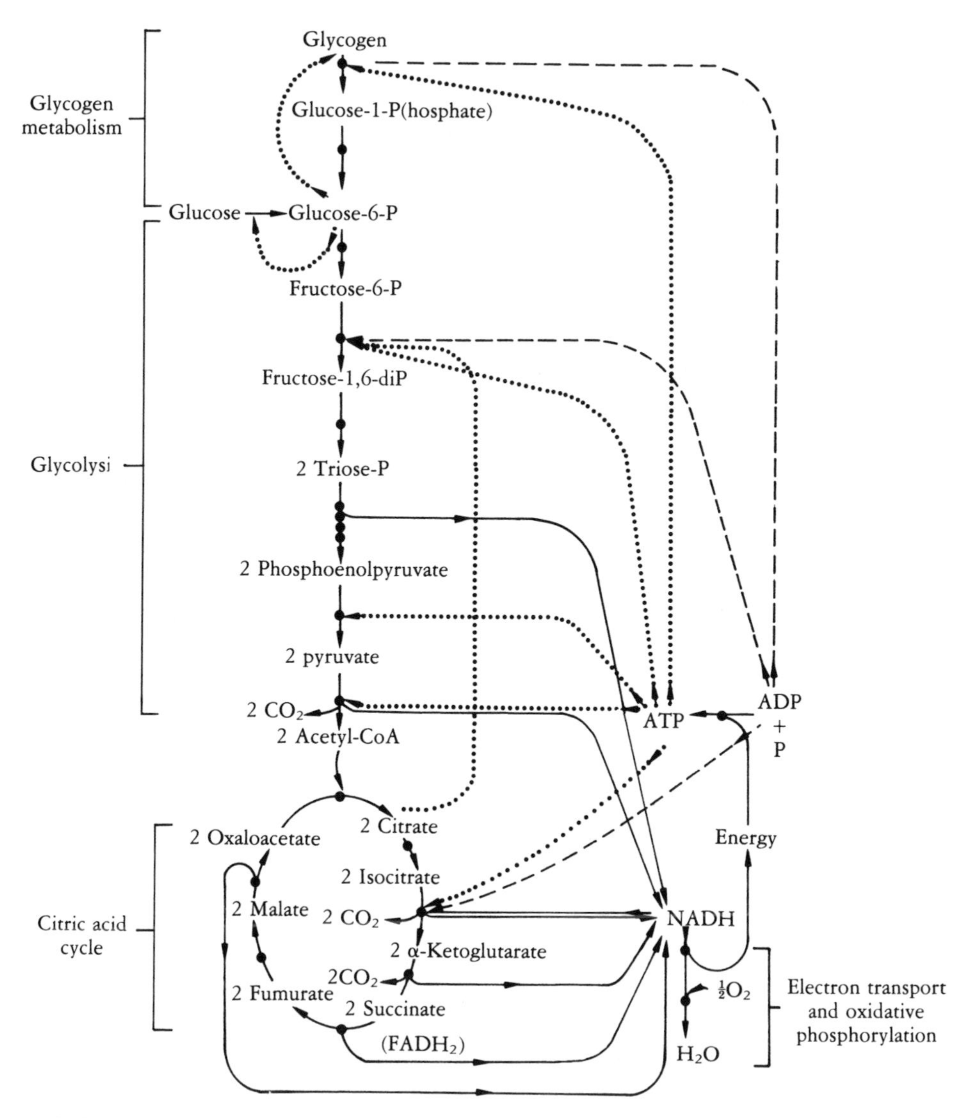


**FIGURE 3-1**
Control of ATP generation. Dotted lines denote inhibitory feedbacks, and dashed lines, excitatory feedbacks.

by ATP before their polymerization into proteins. Both the glycosidic linkages in glycogen and the peptide bonds in proteins are dehydration linkages and therefore require energy in an aqueous milieu.

The active form of the enzyme glycogen synthase, or synthase I, catalyzes the polymerization of glucose monomers into glycogen. Another enzyme, synthase I kinase, can phosphorylate synthase I, using ATP, to produce synthase D, an enzyme that is a much less active form of synthase I. Still another enzyme,

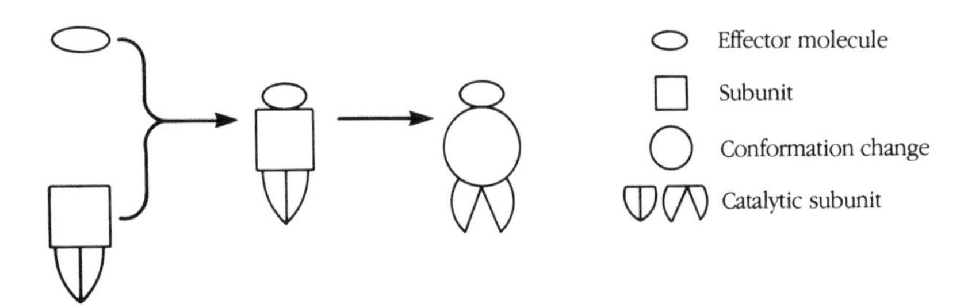


**FIGURE 3-2**
Allosteric regulation. The effector molecule (⬭) binds one subunit (□), causing a conformational change (○), which is then communicated to the catalytic subunit.

synthase phosphatase, removes the phosphate on synthase D, thereby regenerating synthase I.

An overabundance of glycogen inhibits synthase phosphatase. This means that most of the synthase I will end up as synthase D, which only weakly catalyzes glycogen synthesis. The synthase I kinase is, in turn, allosterically regulated by cyclic AMP: It has its active form when cyclic AMP is bound and its inactive form when cyclic AMP is absent. Cyclic AMP, in turn, is made

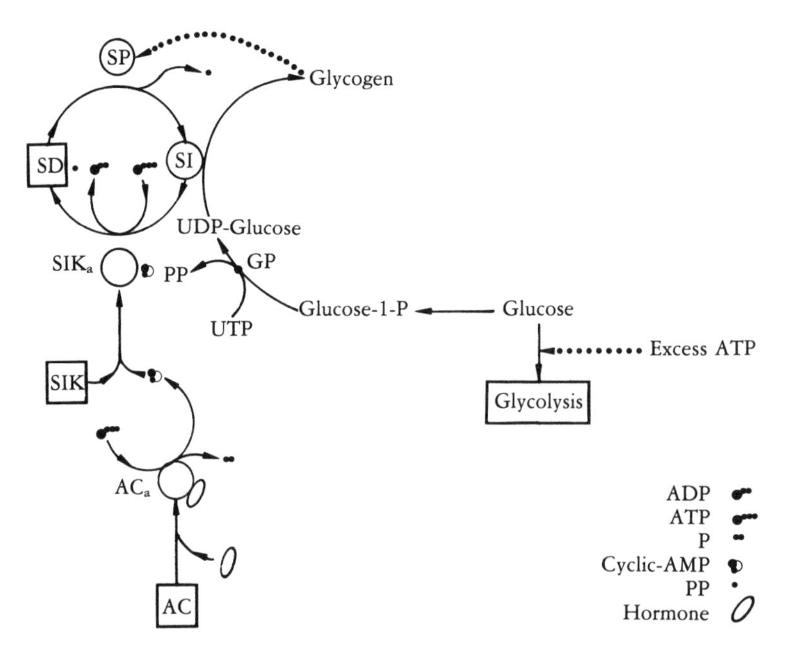


**FIGURE 3-3**
Glycogen synthesis control. SI denotes synthase I; SD, synthase D; SIK and $SIK_a$, the inactive and active forms of synthase I kinase, respectively; AC and $AC_a$, the inactive and active forms of adenyl cyclase, respectively; SP, synthase phosphatase; and GP, UDP-glucose pyrophosphorylase. Dotted lines denote the inhibitory effect of glycogen on SP and of excess ATP on glycolysis.

from ATP by adenyl cyclase, an enzyme that is regulated by such hormones as epinephrine and glucagon. Note the variety of ways that phosphate is involved in this control mechanism!

***Glycogen Breakdown*** Although the process of glycogen synthesis is complex, it involves only one kinase and one phosphatase. The degradation (catabolism) of glycogen to glucose-1-phosphate is more elaborate (see Figure 3-4), involving two kinases and two phosphatases. The sequence begins as a phosphorolysis, in which the enzyme phosphorylase a splits one glucose monomer from the glycogen polymer, forming glucose-1-phosphate. Phosphorylase a is a tetramer; each of its four subunits contains a phosphorylated serine residue. The enzyme phosphorylase phosphatase removes these phosphates, converting the tetrameric enzyme complex into two dimeric molecules of phosphorylase b, which is the inactive form of phosphorylase a. Rephosphorylation of the serine residues causes the tetrameric complex to reassemble, restoring the active form of the enzyme. This phosphorylation is catalyzed by phosphorylase b kinase, which also has active and inactive forms. The active form of phosphorylase b kinase results from phosphorylation of its inactive form, with ATP as phosphate donor in a reaction catalyzed by phosphorylase

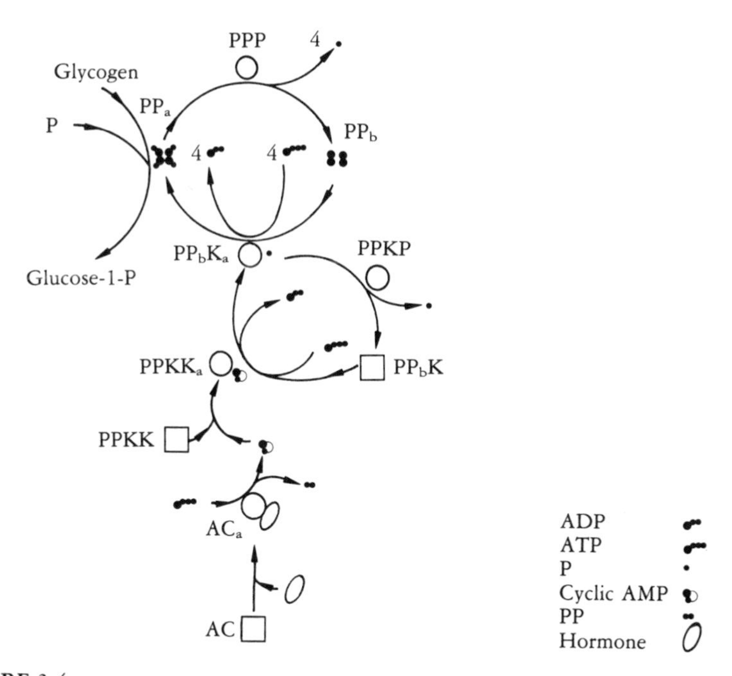


**FIGURE 3-4**
Glycogen utilization. PP denotes phosphorylase a; PP$_b$, phosphorylase b; PPP, phosphorylase phosphatase; PP$_b$K$_a$, phosphorylase b kinase in active form; PP$_b$K, phosphorylase b kinase in inactive form; PPKP, phosphorylase kinase phosphatase; PPKK$_a$, phosphorylase kinase kinase in active form; PPKK, phosphorylase kinase kinase in inactive form; AC, adenyl cyclase in inactive form; and AC$_a$, the active form of adenyl cyclase.

kinase kinase. This enzyme also has active and inactive forms. The active form results when cyclic AMP binds to the inactive form. Cyclic AMP is made, as before, from ATP by the catalytic action of adenyl cyclase, which changes from its inactive form to its active form after binding hormones, such as epinephrine and glucagon. In the synthase system the presence of these hormones ultimately causes decreased glycogen synthesis, whereas in the phosphorylase system the presence of hormones ultimately activates the catabolism of glycogen; both tendencies lead to an increase in the amount of glucose. The effect of this cascade of activations is that a small amount of hormone leads to a large amount of glycogen catabolism, as if the system amplified the initial hormonal signal.

Hormones usually bind to the outsides of cells, and the cyclic AMP (produced by the activation of adenyl cyclase) is on the inside. Adenyl cyclase (depicted as $AC_a$ in Figures 3-3 and 3-4) is a complex that usually has an outer subunit that binds specific hormones and an inner subunit on which the cyclase activity resides. This allosteric activation of adenyl cyclase constitutes the *second-messenger* mechanism of hormone regulation.

### Amino Acid Synthesis

Allosteric feedback also controls amino-acid synthesis. A sufficient concentration of an amino acid inhibits an enzyme that acts near the beginning of the synthesis pathway. Some pathways have steps that are the same for several amino acids; these steps are catalyzed by enzymes that occur as a set of variations called isozymes. For example, a transaminase enzyme that uses pyridoxal phosphate as a coenzyme and glutamic acid as a source for an amino group converts the intermediate of the tricarboxylic acid cycle, oxaloacetic acid, directly into aspartic acid. Aspartic acid, in turn, is converted into lysine, methionine, and threonine (Figure 3-5). The initial step is a phosphorylation that

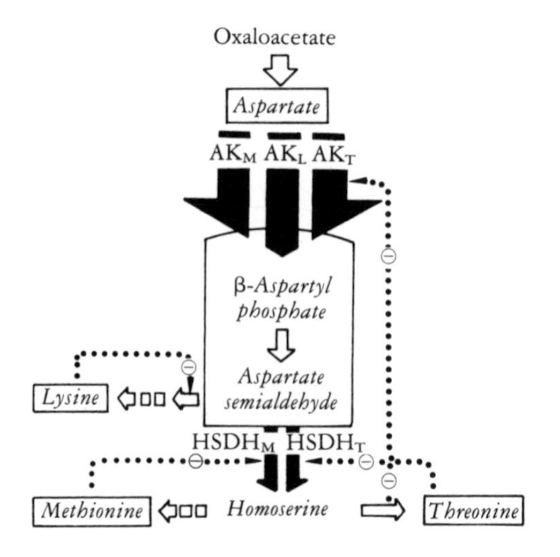


FIGURE 3-5
Amino acid regulation. Dotted lines denote allosteric inhibitory feedback; and broken arrows, successive intermediate steps not explicitly depicted.

is catalyzed by aspartate kinase, which has three isozymes—$AK_M$, $AK_L$, and $AK_T$—whose syntheses are repressed by methionine, lysine, and threonine, respectively. Further along the pathway, aspartate semialdehyde is converted into homoserine by two isozymes of homoserine dehydrogenase, $HSDH_M$ and $HSDH_T$, whose syntheses are repressed, respectively, by methionine and threonine. Two levels of regulation are therefore indicated: the synthesis of an enzyme and the allosteric switching on and off of an already synthesized enzyme. The syntheses of other amino acid groups are regulated by similar dual mechanisms.

The point of emphasis is that key steps in these reactions involve phosphate. Because so many contemporary control mechanisms in highly evolved organisms use phosphate, it must have played a role as a regulatory substance in ancestral species, even the primitive uroboros.

## 3-3 PHOSPHATE MECHANISMS IN ORGANISMS

Not only does phosphate play the central role in molecular energy transactions, but it also plays a key role in many kinds of cellular control processes. Phosphorylation (and dephosphorylation) is a widespread mechanism of regulation. The phosphorylation steps are catalyzed by kinases, which are often controlled by cyclic-AMP, another phosphate derivative. If the assumption that phosphate is especially well suited for such processes is correct, then we would expect to see such mechanisms exhibited in higher (multicellular) forms of life.

### Examples of Contemporary Control Mechanisms

Diverse but representative examples of control mechanisms in contemporary organisms follow. Many more examples are possible and can be found in cell biology textbooks. Again, some readers may prefer to skip this material and pick up the general discussion near the end of this section.

*Message Transmission in Animals* The chief mechanism of message transmission between cells in the nervous systems of animals is the use of cyclic AMP-activated kinase. This mechanism exists throughout the nervous systems and in hormone-receptors of many species. The structure is shown in Figure 3-6.

Dissociation of this enzyme system results in inactive proteins. However, researchers have had some success in getting the system to spontaneously reassemble in artificial phospholipid vesicles. Sulfur, in the form of an —SH group, plays a role in the G-C interaction, and $Mg^{2+}$ is also required. Some evidence exists that the amino acid sequence of the G protein is related to the comparable GTP-binding-protein sequence of the protein biosynthesis machinery. These features may be clues to the evolution of this mechanism.

The message-transmission mechanism just described is capable of a great variety of specific functions, partly as a result of the many hormones or hor-

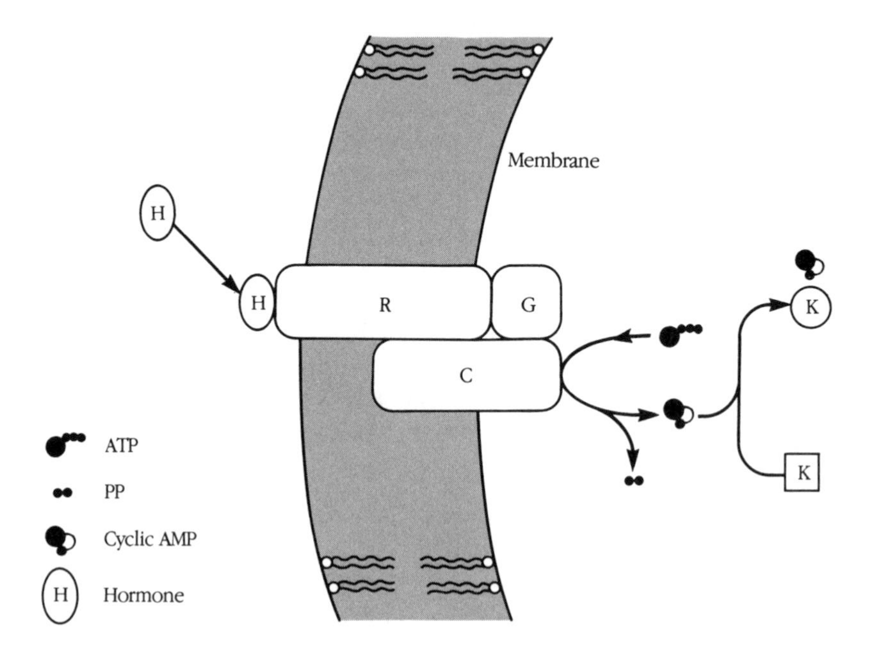


**FIGURE 3-6**
A hormone H (or neurotransmitter) binds to a receptor protein R on the outside of the target cell. This induces a conformational change in R which is communicated to the protein G through weak bonds. Initially, guanine-diphosphate (GDP) was bound to G; but the conformational change in R induces the release of GDP and the binding of GTP. Now R can release the hormone H, which enzymes may degrade, and return to its initial state. As long as GTP is not hydrolyzed to GDP and P, protein G induces a conformational change in protein C, the enzyme that makes cyclic-AMP from ATP. This occurs inside the cell. Cyclic-AMP then binds to inactive protein kinase K, making it active $K_a$, which in turn catalyzes phosphorylation of a target enzyme system. When the GTP hydrolyzes, the mechanism must be restimulated by hormone. In general, the target system becomes either excited or inhibited upon phosphorylation by the kinase, depending on the specific case.

monelike molecules, such as neurotransmitters, that function as *first messenger*, including dopamine, serotonin, norepinephrine, acetylcholine, insulin, interferon, opiates, steroids, and nerve growth factor. Variations of the above mechanism that use something other than cyclic-AMP as second messenger, such as cyclic-GMP or calcium, also add functional diversity. Furthermore, it is not always a kinase that is activated by the second messenger, but sometimes a phosphatase. Perhaps these variations evolved from the less diverse phosphate mechanism of the primitive uroboros.

Some of these systems are located throughout the brain, whereas others are partically or highly localized. Some are as complex as the glycogen degradation system, and some involve interactions with related systems. Regulations of these systems can occur at almost any step. For example, the mechanism of action of the cyclic-AMP-dependent kinase is the phosphorylation of the target protein using ATP as donor.

The message-transmission mechanism is reversible because the cyclic nucleotides are metabolized by phosphodiester isozymes, and phosphatases dephosphorylate target proteins. All these processes of activation and inactivation, of phosphorylation and dephosphorylation, provide a controlled balance of metabolites and processes in the cell. Although their utility is obvious, their evolution is not.

*Muscle Function and Control* The two major components of muscle protein are myosin and actin, which form a complex with several other proteins. Cross bridges form between actin and myosin filaments (de Duve, 1984; Harold, 1986). The regulation of smooth muscle contraction by $Ca^{2+}$ and cyclic-AMP involves regulation of myosin phosphorylation. Adenyl cyclase is present, as well as three kinases and two phosphatases. Calcium binds the protein *calmodulin*, which in turn stimulates or inhibits a kinase, depending on which aspect of the system is involved (Figure 3-7).

A nerve pulse to a neuromuscular junction depolarizes the muscle mem-

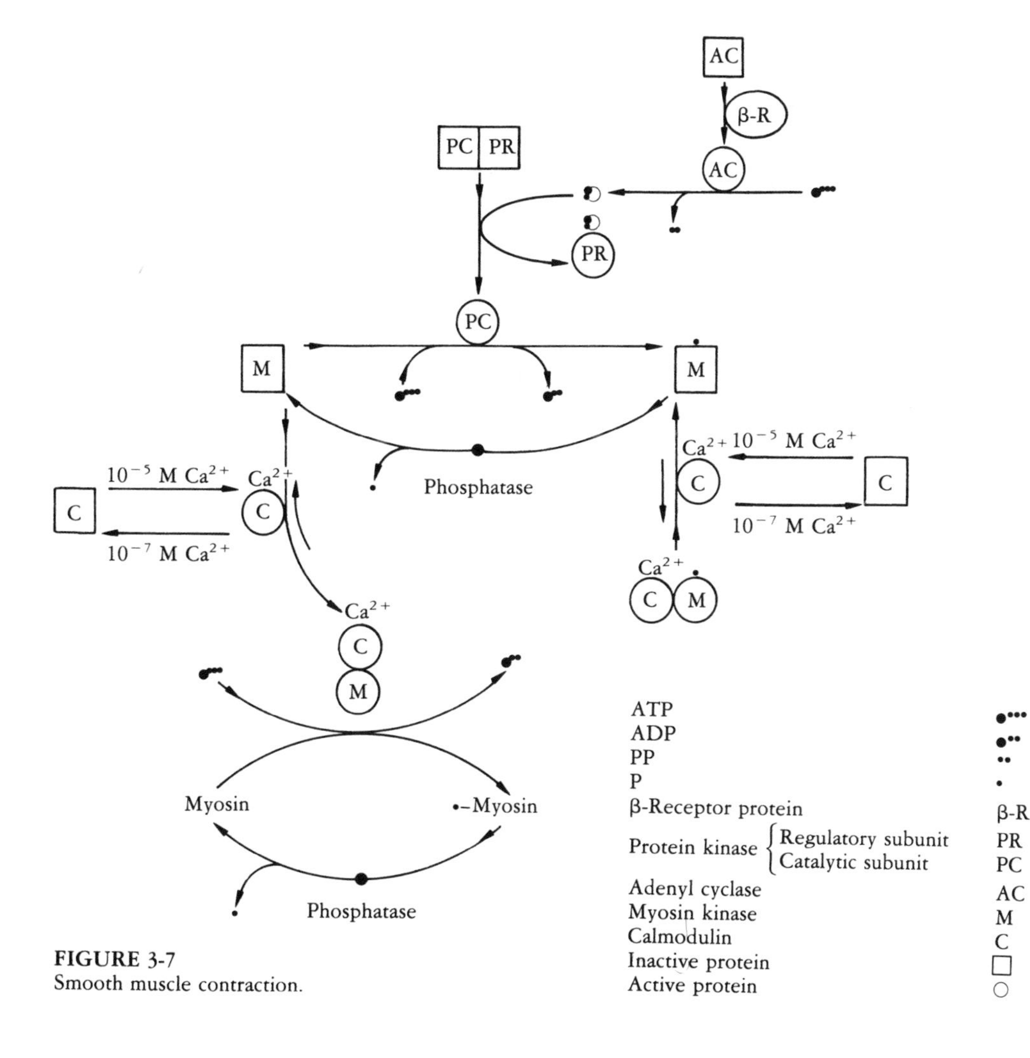


FIGURE 3-7
Smooth muscle contraction.

brane, leading to an increase in ion permeability. Calcium activates myosin's hydrolysis of ATP, the energy source for muscle action. This ATPase activity of myosin breaks the cross links between actin and myosin. When the cross links subsequently reform, the relative positions of actin and myosin shift, which is manifested as muscle contraction. Myosin-based regulation occurs in mollusca muscle, in vertebrate smooth muscle, and even in nonmuscle cells, containing myosin. However, in vertebrate skeletal muscles and in cardiac muscle, the control mechanism is actin-based.

Note that both phosphates and calcium play significant roles.

***Calmodulin*** The calcium-binding protein, calmodulin, is prevalent in organisms. The $Ca^{2+}$-calmodulin complex can act directly in $Ca^{2+}$ transport processes, as in the ATPase system, or indirectly on a regulatory system through a protein kinase, as described above. Calmodulin is not tissue or species specific, and its amino acid sequence has been extremely well conserved during evolution. In these respects, calcium plays a role paralleling the role of cyclic-AMP in regulation.

Calmodulin is a protein with a molecular weight of about 17,000. It contains no tryptophan or cystein but has an abundance (about 50 percent by weight) of glutamic and aspartic acid residues, which is reminiscent of proteinoids (see Table 2-1). It has been found in many invertebrates, protozoa, and in vertebrates. These facts suggest that calmodulin is a primitive protein like the iron-sulfur proteins discussed in Chapter 2. In vertebrates, it is most highly concentrated in brain and testes. It has a role in myosin kinase regulation. In other cells, including unicellular organisms, it helps to disassemble microtubules, which are part of the cytoskeleton of cells and are involved in cell motility, chromosome motion during mitosis, and axonal transport. The full range of calmodulin functions and their mechanisms is not adequately known.

***Histone Regulation*** Phosphorylation and kinases also regulate histone binding to DNA in eukaryote chromosomes. Histones are the major class of nuclear proteins, that is, DNA binding proteins. Phosphorylation alters their binding to DNA and has been demonstrated to be necessary, although not sufficient, for gene activity (transcription or replication). In a few cases researchers have demonstrated that the phosphorylation is triggered by cyclic-AMP and that histone phosphorylation appears to be the target for virus-induced tumor kinase activity. Similar mechanisms appear to be involved in *E. coli* infections by bacteriophage T7, indicating that such gene expression processes (cyclic-AMP triggered phosphorylations) occur in prokaryotes.

***Slime Mold Development*** This example of phosphate-dependent control describes the differentiation and development of the slime mold, *Dictyostelium discoideum*. which is a model system for differentiation and development in higher systems. The life cycle of the slime mold has two distinct stages, a vegetative stage as independent amoeboid cells and a generative stage as a multicellular organism. Nutrient deprivation triggers the transition from unicellular to multicellular existence: All cells release into the medium a diffusible substance that attracts other cells. The cells move together and aggregate into multicellular groups containing $10^5$ individuals. These aggregations then differentiate, eventually producing spores that give rise to unicellular amoebae.

The attractor substance is cyclic-AMP, made from ATP by adenyl cyclase. Cyclic-GMP is also involved. Cyclic-AMP is bound to receptors on the outsides of the cells. $Ca^{2+}$ appears to be necessary for the cyclase activity. The movement of the cells that cyclic-AMP induces appears to be mediated by actin and myosin. At least 13 protein kinases have been implicated in these processes. Histone H2A is one target of kinase, and myosin is another. Investigators are still studying the sequence of this activity and the consequences, but it is already clear that kinases play the major role in cell movements. To see $Ca^{2+}$, actin and myosin, and cyclic-AMP involved in cell motion in this manner suggests how true multicellular organisms may have evolved muscle.

## Primitive Control Mechanisms: Speculations

Living systems have clearly used protein phosphorylation as a regulatory and developmental device throughout evolution. Evolutionary insight into these processes will come from the techniques of protein and polynucleotide sequencing, which permit the establishment of lineages and kinships. This approach has been fruitful in a number of cases already, such as cytochromes, ferredoxins, and calmodulin.

It is already clear that phosphate is a principal component, in one form or another, in regulatory mechanisms. ATP is the ultimate source of phosphate in most phosphorylation mechanisms and is usually directly involved in the reactions. Earlier, I proposed that pyrophosphate may have been its evolutionary precursor in energy metabolism. Because research shows that pyrophosphate can substitute for ATP in some reactions, including kinease activity, it could have served as a donor of phosphate in regulatory processes even at the level of a protenoid system, such as the primitive uroboros.

***Pyrophosphate in Primordial Regulation*** I have developed a view in which oxidative phosphorylation or chemiosmosis converted environmentally abundant oxidation-reduction energy into abundant phosphate bond energy in pyrophosphate. This energy, instead of simply being hydrolyzed directly into heat, drove polymerization and then, through phosphorylation, engaged in the regulation of metabolic functions that took place in self-assembled aggregates of these polymers. I have rationalized this direction of energy flow by saying that the outcome is dictated by rates of reactions. Earlier I provided a conceptual scheme for how this sequence could occur based upon pyrophosphate-powered proteinoid microspheres, but another question remains: Is the observed outcome, that is, the existence of living matter, the result of emergent properties or does it reflect some underlying principle, hitherto unenunciated?

### Variational Principles

Physicists are fond of reducing the basic laws of physics to extremal, or variational, principles. For example, Hamilton's principle of least action subsumes Newton's laws of mechanics; Rayleigh's principle of least dissipation (of Joule heating in resistances) subsumes Kirchhoff's electrical circuit laws; and an analog of Rayleigh's principle of least dissipation can succinctly express even Onsager's theory of near-equilibrium, nonequilibrium thermodynamics. Is there an analogous underlying principle at work in the molecular biology of phosphate bond energy use?

In the context of electrical circuits, Rayleigh's least-dissipation principle says that given two circuit pathways, an electrical current will flow primarily along the pathway that produces the lesser Joule heating, that is, the path of least (electrical) resistance. Perhaps an analogous principle directs phosphate bond energy use away from heat-producing hydrolysis and toward monomer activation, synthesis, and regulation. More likely, a principle of least-dissipation of Gibbs free energy is closer to the truth.

### Why Phosphorus?

Why does the story of biological energy use revolve around phosphorus rather than, say, nitrogen or sulfur? Why is calcium used as a second messenger, in parallel with cyclic-AMP? Why not magnesium instead? Why are organic molecules based on carbon instead of silicon? Researchers have asked many such questions; in some cases, they even have answers. The tetravalency and relative stability of carbon bonds are responsible for the vast array of organic compounds. And the electronic structures of the elements determine many of their properties, including the suitability of phosphorus for energy metabolism. However, given the primordial dozen elements, many of the observed characteristics of the living state *cannot be predicted from first principles*. The current description of the living state is largely ex post facto and is highlighted by emergent properties.

## 3-4. THE EVOLUTION OF ENERGY METABOLISM AND STORAGE

Major transitions in form and behavior have occurred in the course of the evolution of life on earth: transitions from unicellular to multicellar life, from primitive heterotroph to autotroph to modern heterotroph, from anaerobe to photoautotroph to aerobe, from sessile organism to motile, and from aquatic organism to terrestrial. (See the geological time scale, Table 2-3.) These transitions are so well delineated that this section will consider them in terms of metamorphosis at the phylogenetic level, rather than in the usual biological context of ontogeny. In every instance, the metamorphosis is directly coupled to a significant change in energy metabolism, regulation, and storage. I hypothesize that *advances in energy metabolism, regulation, and storage have been the impetus to phylogenetic metamorphosis*. As a corollary, transitions in energy metabolism, regulation, and storage predicate ontological metamorphosis as well.

I shall examine this hypothesis from the long-term perspective of phylogenetic changes, particularly the emergence of multicellular motility (that is, muscle) and higher forms of nervous systems. The focus is on excitable tissues, muscle and nerve, and their requirement for energy storage molecules capable of sustaining rapid energy utilization. The discussion will show that both phylogenetic and embryological considerations strongly support the hypothesis.

## Phosphagens

Phosphagens are energy storage molecules that occur throughout animal phyla and in protists. They differ from energy storage molecules such as glycogen in that they provide immediate energy when it is needed. Phosphagens are of three major types: polyphosphates, arginine phosphate (Arg—P), and creatine phosphate (Cr—P).

***Phosphagen Function*** ATP/ADP ratios regulate many metabolic pathways (see Figure 3-1). Recall that ATP abundance tends to inhibit energy-yielding pathways, whereas ADP abundance tends to activate them. However, when an organism has an abundant supply of convertible energy in a food source, such as carbohydrate, it is not advantageous for the organism to transform some of the food energy into ATP and then shut down the process because of raised levels of unused ATP. It is better for the organism to convert ATP into a storage form of phosphate, so as to maintain reasonable ATP/ADP ratios, and thereby process all the available food source. This, presumably, is what polyphosphate synthesis and storage accomplishes in the prokaryote energy economy.

In multicellular eukaryotes, the purpose of the phosphagens is clear: Arg—P and Cr—P are energy storage molecules that are quickly available for use in phosphorolyzing ADP into ATP.

***Phylogenetic Distribution of Phosphagens*** Researchers have found large amounts of polyphosphates in protozoa in the form of granules that are commonly associated with DNA fibrils and cellular zones rich in ribosomes. In unicellular eukaryotes, such as yeast and fungi, polyphosphates may constitute 20 percent of the dry weight. In both prokaryotes and unicellular eukaryotes, polyphosphates serve several functions, but evidence exists in both cases for their function as ready sources of high-energy phosphate—that is, as phosphagens. Only micoorganisms use polyphosphates as phosphagens, whereas multicellular animals use Arg—P and Cr—P. The difference between Arg—P users and Cr—P users is illuminating. Superficially it appears that invertebrates use Arg—P and vertebrates use Cr—P, but closer inspection of the evidence shows a sharper distinction: Prechordata use Arg—P, and Chordata use Cr—P. The Chordata include all vertebrates and the so-called Protochordata. Protochordata comprise three classes: Tunicata (sea-squirts), Enteropneusta (acorn worms), and Cephalochorda (lancelets or amphioxus). Acorn worms, the most primitive invertebrate chordates, have a nervous system that is the most primitive of any group of animals having an organ-system level of construction.

Three anatomical structures distinguish Cr—P users: (1) a notochord, a cartilagenous rod extending the length of the body and supporting the soft tissues; (2) a dorsal tubular nerve cord; and (3) pharyngeal gill slits. Embryos of all Protochordata and Chordata posses these three structures. The Prechordata, which do not possess these structures, include all other multicellular invertebrates, such as Coelenterata, Platyhelminthes, Nemertea, Mollusca, Arthropoda, and Echinodermata.

Biochemical analyses of many phyla and species seem to indicate that Arg—P, as a source of quick energy, supports muscle activity and rudimentary nervous activity, whereas Cr—P, which also supports muscle activity, is essential for higher level nervous tissue activity. The notion that these phosphagens are *prerequisites* to muscle and nerve activity follows from their functions: The energy demand of muscle is so great that glycolysis and electron transport cannot regenerate ATP fast enough to supply it. Mammalian muscle can work at a rate that demands $10^{-3}$ moles of ATP per gram of muscle per minute. The ATP available per gram of resting muscle is $5 \times 10^{-6}$ mol. This is sufficient for only 0.3 s of activity. The phosphagen stores, however, provide a rapidly regenerated source of ATP that can last for several minutes.

The evolutionary perspective is that early multicellular species, probably worms, were able to use phosphagens (originally stored merely to maintain ATP/ADP ratios while accumulating abundant energy from food sources) to energize muscle tissue for rapid motility. An evolutionary relationship appears to exist among the muscle proteins (actin and myosin) and related proteins that confer a cytoskeleton and motility on unicellular forms. Muscle seems a natural development in multicellular forms, but its use is possible only if phosphagens are available. The sequence hypothesized here is that abundant energy in the environment leads to energy storage in phosphagens, which then leads to cytoskeletons and simple, cell movement and to the appearance of muscle tissue, both of which conferring evolutionary advantages.

***Embryological Considerations*** Embryology strengthens the preceding views. Motility, which began in unicellular forms, plays a crucial role in the development of a multicellular organism, especially during the early stages of embryogenesis when blastula and gastrula form. Only two lines of embryological development exist in evolution: the arthropod line and the chordate line. These two lines differ in the way that mesoderm cells form in the gastrula stage (Figure 3-8). In the arthropod line (including flatworms, nemerteans, mollusks, annelids, and arthropods), the primitive mesoderm cells bud off from primitive endoderm cells of the blastula. In the chordate line (including all Chordata and the Echinodermata), the mesoderm cells bud off as coelomic sacs from the endoderm.

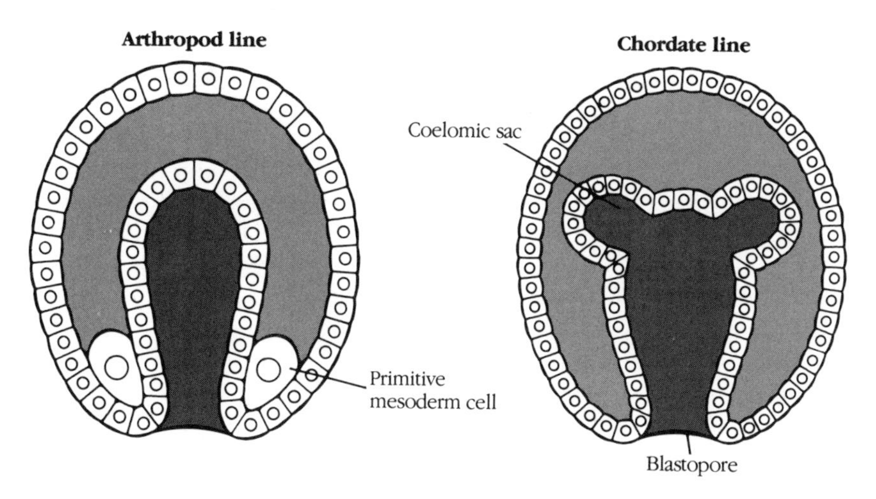


**FIGURE 3-8**
Arthropod and chordate mesodern cells.

The Echinodermata, which are invertebrates, might appear to be in the "wrong" lineage, but evolutionists view them as the link between other invertebrates and the Chordata. Embryologically, they are so similar to acorn worms that an early specialist on echinoderms described the larva of an acorn worm as an echinoderm larva. However, the acorn worm develops into a true Protochordata, whereas the echinoderm larva ends up as a starfish or sea urchin, without a sophisticated nervous system. Biochemically, echinoderms use both Arg—P and Cr—P as phosphagens; thus they are transitional organisms from the phosphagen viewpoint as well.

The transition from Arg—P to Cr—P correlates with dramatic differences in the embryogenesis of the various gastrular cell types: mesoderm cells, which give rise to muscles and vascular system; ectoderm cells, which give rise to skin and nervous system; and endoderm cells, which give rise to alimentary canal and digestive glands. Thus their fates determine almost the entire structure of the organism. Bones derive mostly from neural crest cells, which begin to develop after the neural plate has formed from ectoderm cells.

These developmental events require cell motility within the developing gastrula and this requires phosphagen. Why Cr—P leads to one outcome and Arg—P to another is unknown. Nevertheless, with Cr—P, sophisticated nervous tissue, protected by bone, evolved.

***Evolutionary Refinements and Chemistry of Phosphagens*** About the time that the chordate and the arthropod lines separated from a common ancestor, only wormlike organisms made up the arthropod line: true arthropods and mollusks arose later from ancestral worms. The trigger for this separation into two great lines may have been the emergence in chordates of Cr—P in place of Arg—P. At this time, evolution may have tried other phosphagens as well.

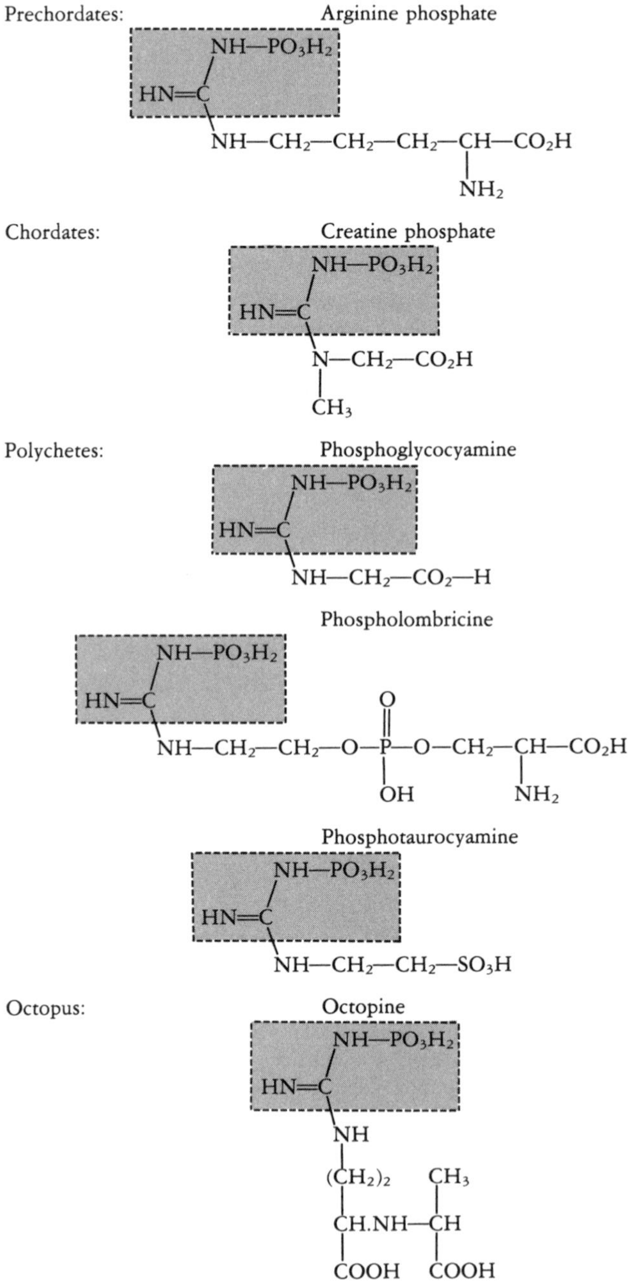

Prechordates: Arginine phosphate

$$\mathrm{HN{=}C(NH{-}PO_3H_2){-}NH{-}CH_2{-}CH_2{-}CH_2{-}CH(NH_2){-}CO_2H}$$

Chordates: Creatine phosphate

$$\mathrm{HN{=}C(NH{-}PO_3H_2){-}N(CH_3){-}CH_2{-}CO_2H}$$

Polychetes: Phosphoglycocyamine

$$\mathrm{HN{=}C(NH{-}PO_3H_2){-}NH{-}CH_2{-}CO_2{-}H}$$

Phospholombricine

$$\mathrm{HN{=}C(NH{-}PO_3H_2){-}NH{-}CH_2{-}CH_2{-}O{-}P({=}O)(OH){-}O{-}CH_2{-}CH(NH_2){-}CO_2H}$$

Phosphotaurocyamine

$$\mathrm{HN{=}C(NH{-}PO_3H_2){-}NH{-}CH_2{-}CH_2{-}SO_3H}$$

Octopus: Octopine

$$\mathrm{HN{=}C(NH{-}PO_3H_2){-}NH{-}(CH_2)_2{-}CH(COOH).NH{-}CH(CH_3)(COOH)}$$

**FIGURE 3-9**
Phosphagens using the energy-rich —N~P bond.

And, in fact, researchers have found several other phosphagens in the annelid worms called polychetes. Even the octopus, a mollusk, has its own unique phosphagen. Chemically, all these phosphagens are similar to Arg—P and Cr—P (Figure 3-9), and each involves the energy rich —N~P bond. Such compounds are effective donors of energy-rich phosphate for conversion of ADP to ATP.

### Skeletal Structure Correlates

The cytoskeleton of cells is based on proteins that are related, or in some cases identical, to the muscle proteins actin and myosin. Their energy comes from ATP, and cyclic-AMP and calcium trigger the kinases that regulate their action. The early multicellular life forms were worms (flat worms), which achieved advantageous motility with their muscles because they stored energy as Arg—P. Calcium is a component of the worm's muscle regulatory processes.

In the arthropod lineage, the earliest mollusks that evolved were little more than worms with protective shells of calcium carbonate, true mollusks evolved later. Mollusk shells have formed deposits of limestone nearly everywhere on earth. This calcium carbonate appears to be a natural product of a calcium-controlled energetic process coupled to a metabolism that produces the waste product $CO_2$, because these organisms are aerobes which produce $CO_2$ as an end product of energy metabolism.

The chordate lineage uses Cr—P, which may have provided an enhanced rate of phosphate utilization. Not only did a more sophisticated nervous system develop, apparently correlated with Cr—P, but nervous tissue used more cal-

**TABLE 3-1**
*Mineralized Tissues*

| *Species* | *Tissue* | *Mineral crystalline form* | *Organic matrix* |
|---|---|---|---|
| Plants | Cell walls | Calcite ($CaCO_3$) | Cellulose, pectins, lignins |
| Diatoms | Exoskeleton | Silica ($SiO_2$) | Pectins |
| Mollusks | Exoskeleton | Calcite, aragonite ($CaCO_3$) | Protein (conchiolin) |
| Arthropods | Exoskeleton | Calcite ($CaCO_3$) | Chitin, proteins |
| Vertebrates | Endoskeleton, bone, cartilage, tooth | Hydroxyapatite ($Ca_{10}(PO_4)_6(OH)_2$) | Collagen |

cium and phosphate, both as control substances ($Ca^{2+}$ and phosphate) and as active ion fluxes ($Ca^{2+}$). Muscle contains substrate-level amounts of both $Ca^{2+}$ and phosphate because the protein filaments of actin and myosin have $Ca^{2+}$ and phosphate sites about every 400 Å along their lengths. Thus the deposition of calcium phosphate as bone seems to be a natural consequence of chordate metabolism. Calcium phosphate is a stronger material than calcium carbonate. Table 3-1 summarizes the occurrences of skeletal structures in organisms.

## REFERENCES

**Section 3-1**

Fox, R. F., *Biological Energy Transduction: The Uroboros,* John Wiley, New York, 1982.

Nomura, M., "Assembly of Bacterial Ribosomes," *Science* 179 (1973): 864.

Schrodinger, E., *What is Life?,* Cambridge University Press, Cambridge, England, 1944.

**Section 3-2**

Rindt, K. P., "Expression of Isozymes and Their Function in Differentiation," in *Cell Differentiation,* edited by L. Nover, M. Luckner, and B. Parthier, Springer-Verlag, Berlin, 1982.

**Section 3-3**

Adelstein, R. S., and E. Eisenberg, *Annual Reviews of Biochemistry* 49 (1980): 921.

deDuve, C., *A Guided Tour of the Living Cell,* W. H. Freeman, New York, 1984, Chapter 12.

Eisenberg, E., and T. L. Hill, "Muscle Contraction and Free Energy Transduction in Biological Systems," *Science* 227, (1985): 999–1006. This paper contains a detailed account of how the muscle proteins, actin and myosin, are arranged in space and how ATP causes contraction of the muscle.

Harold, F. M., *The Vital Force: A Study of Bioenergetics,* W. H. Freeman, New York, 1986, Chapter 11.

Keizer, J., "Variational Principles in Nonequilibrium Thermodynamics," *BioSystems* 8 (1977): 219–226.

Klee, C. B., T. H. Crouch, and P. G. Richman, *Annual Reviews of Biochemistry* 49 (1980): 489.

Rubin, C. S., and O. M. Rosen, "Protein Phosphorylation," *Annual Reviews of Biochemistry* 44 (1975): 831.

Schramm, M., and Z. Selinger, *Science* 225 (1984): 1350.

**Section 3-4**

Baldwin, E., *Dynamic Aspects of Biochemistry,* 4th edition, Cambridge University Press, Cambridge, England, 1965.

Buchsbaum, R., *Animals without Backbones,* University of Chicago Press, Chicago, 1948.

Thoai, N. V., and Y. Robin, "Distribution of Phosphagens in Errant and Sedentary Polychaeta," in *Studies in Comparative Biochemistry,* edited by K. A. Munday, Pergamon Press, New York, 1965.

White, A., P. Handler, and E. L. Smith, *Principles of Biochemistry,* 3rd edition, McGraw-Hill, New York, 1964, pp. 749–751.

# CHAPTER 4

# *Nonlinear Dynamics*

The mathematical character of driven, nonlinear dissipative systems is the main topic of this chapter. The important feature of nonlinear dynamics for this discussion is that the equations used to describe such systems do not usually possess closed-form solutions and must be computer modeled. Computer simulation demonstrates that even very simple equations of this type can show extremely complex and sometimes chaotic behavior. The behaviors of such systems are reminiscent of the behaviors of living systems; thus studies of such systems provide insight into life's evolution.

The chapter begins with an example: Turing's mathematical approach to the description of the embryological formation and differentiation of tissue (Section 4-2). This is followed by discussions of mathematical dimension (Section 4-3), numerical solutions for a simple Turing model (Section 4-4), and properties of chaotic attractors (Section 4-5). Harmonic oscillator and damped pendulum systems demonstrate how the behavior of differential equations can change with dissipative perturbations (Section 4-6). Section 4-7 shows examples of emergent behavior and dynamic chaos, and Sections 4-8 and 4-9 discuss models that demonstrate nonlinear dynamics, especially in the matter of predictability. Chapter 4-10 describes the usefulness to organisms of rapid simulation and concludes that the evolutionary significance of the brain is that it has become a rapid simulator of nonlinear processes.

Mathematical terminology and mechanisms abound in Sections 4-2 through 4-9. Readers who feel uncomfortable with the mathematics can use these sections for later reference and go on to Section 4-10. However, anyone who has a rudimentary programming ability (or a friend who does) could easily work through Section 4-9.

## 4-1. THE PHYSICS AND MATHEMATICS OF ENERGY-DRIVEN SYSTEMS

Energy flow is a necessary but not a sufficient condition for the living state of matter since a living state also evolves. The conditions of energy flow and capacity to evolve are both satisfied by phosphate energy transduction interacting with proteins and polynucleotides; thus the living state is as much a consequence of special substances and their emergent properties as it is a consequence of energy flow. The preceding chapters have emphasized the substances and their emergent properties; this chapter emphasizes some general principles that have biological significance. From the viewpoint of physics and mathematics, the study of energy flow is an active research area, with general ideas still evolving. Much of the work on energy-driven systems involves dynamic descriptions given by nonlinear differential equations that cannot be solved in closed form in terms of standard functions. Serious study of such equations became possible only with the advent of powerful, fast computing machines. However, any reader who has access to a microcomputer can explore some of the behavior of driven nonlinear systems using the simple programs given later in this chapter.

## 4-2. REACTION-DIFFUSION EQUATIONS FOR MORPHOGENESIS

A mathematical approach to understanding embryological morphogenesis (formation and differentiation of tissue) was attempted in 1951 by Alan M. Turing in his classic paper "The Chemical Basis of Morphogenesis" (Turing, 1952).

He proposed that the phenomenon is explained by a system of reaction-diffusion equations, which describe the space and time variations of chemical species as they move through space by diffusion and engage in chemical reactions. Turing's idea was that reaction-diffusion equations govern the space and time movement of *morphogens*, chemical species in a growing embryo that trigger specific developmental events and thereby control its morphology.

### Technical Difficulties with the Mathematics

In general, reaction-diffusion equations are coupled partial differential equations for the concentrations of the substances involved, each concentration variable being a function of three spatial coordinates and one temporal coordinate. Although the diffusion terms are linear in these variables, the reaction terms are generally, at least bilinearly, nonlinear. Therefore, the task of obtaining closed-form solutions of such systems of coupled partial differential equations is formidable at best and usually is impossible. Mathematicians are developing computer simulations for such problems, but these simulations are presently limited by the size of problems that computers can accommodate. A computer simulation must use discrete coordinates to represent the four continuous space and time coordinates, and as few as one hundred discrete values for each of the four coordinates results in $10^8$ discrete coordinate sites, called lattice sites. Every concentration must be known for every one of the $10^8$ sites. Each of these sites is dynamically coupled to the others by a system of difference equations that approximates the system of continuous partial differential equations. Even the most sophisticated modern computers cannot handle this many variables conveniently. Nevertheless, progress in this approach is coming in improved parallel processing and new computer architecture.

### Turing's Model

Turing avoided the difficulty of analyzing coupled partial differential equations by invoking a cellular view; that is, he replaced the continuum by a tissue of cells and replaced partial differential equations in $x$, $y$, $z$, and $t$ by ordinary differential equations in $t$ alone. His motive was the cellular structure of tissues, not the necessities of computers although his approach anticipates such necessities admirably.

For morphogen concentrations $X$ and $Y$, the partial differential equations are

$$\begin{aligned} \frac{\partial}{\partial t} X(\mathbf{r}, t) &= f(X(\mathbf{r}, t)Y(\mathbf{r}, t)) + D\, \nabla^2 X(\mathbf{r}, t) \\ \frac{\partial}{\partial t} Y(\mathbf{r}, t) &= g(X(\mathbf{r}, t)Y(\mathbf{r}, t)) + D'\, \nabla^2 Y(\mathbf{r}, t) \end{aligned} \tag{1}$$

where $D$ and $D'$ are diffusion constants, and $f$ and $g$ are arbitrary, perhaps even nonlinear, functions of the morphogen concentrations. For a ring-shaped assembly of cells, using Turing's cell index $r = 1, 2, \ldots, N$, the system of ordinary differential equations equivalent to Equation (1) is

$$\begin{aligned} \frac{d}{dt} X_r &= f(X_r, Y_r) + \frac{D}{l^2}(X_{r+1} - 2X_r + X_{r-1}) \\ \frac{d}{dt} Y_r &= g(X_r, Y_r) + \frac{D'}{l^2}(Y_{r+1} - 2Y_r + Y_{r-1}) \end{aligned} \tag{2}$$

where $r = 1, 2, \ldots, N$, and $l$ is the center-to-center distance between adjacent cells.

Turing's primary goal was to study the onset of instabilities in the equations, an approach characteristic of many studies that followed. He investigated simple geometric assemblies of cells such as rings, sheets, and spherical shells. He was able to identify six ways for instability to arise in rings. The usual instability was a transition from homogeneity and symmetry to waves and broken symmetry. (The term *instability* is perhaps too strong, both in Turing's work and in later work by others.) What happens is that when the investigator alters a parameter in a certain way, a state of spatial uniformity and time independence is transformed into a time-dependent oscillation of a spatial pattern of some sort. The old state becomes unstable, whereas the new state can be very stable unless the parameter is altered again. Further alterations can lead to new transitions and hence to instability of the prior state and relative stability of a new behavior in space and time. Sequences of such transitions generally occur as a parameter changes. Turing claimed to understand gastrulation by using a spherical shell of cells to represent a blastula, which possesses spherical symmetry. A breakdown of this symmetry occurred under appropriate operating conditions and led to a gastrula stage which has broken symmetry. Turing concluded his analysis by suggesting that digital computers would be useful.

Turing realized that the wave patterns he saw were the result of "a continual supply of free energy." Calling $A$ the substance with greatest free energy and $B$ the substance with least, he wrote that "Energy for the whole process is obtained by the degradation of $A$ into $B$." Substances $C$ are catalysts in his schemes and govern the rates at which the morphogens $X$ and $Y$ change chemically.

Turing's models involved a single mechanism. For appropriate choices of diffusion constants, reaction rates, and initial substance concentrations, an unstable homogeneous spatial state can be a solution to the equations in some situations. For example, perturbations of wavelengths outside some narrow range around a characteristic wavelength $l_0$ decay away, returning to the spatially homogeneous state; or perturbation of wavelengths within the select range around $l_0$ grow, yielding a spatially patterned final state. Stuart A. Kauffman and others (1978) achieved a detailed account of the embryogenesis of the wing imaginal disk of *Drosophila melanogaster* using such a mechanism. In this work, the investigators obtained the closed-form solution of coupled partial differential equations. With appropriate choices for the values of the diffusion constants and reaction rates, the homogeneous spatial state is unstable for thermal perturbations of appropriate wavelength. As a result, spontaneous formation of spatially patterned states occurs. This mechanism predicts the specific pattern of development found for the wing disk when the disk is modeled as a growing ellipse containing two morphogens $X$ and $Y$. The actual chemical identities of $X$ and $Y$ are not required (or known) for this analysis.

Turing's ideas led to the concept of *self-organization* in a spatial system. If conditions are right and instabilities of the type just discussed occur, thermal perturbations favor spatially patterned states over homogeneity. That is, particular circumstances can produce a general result. He fully appreciated that this type of instability occurred in driven systems and that one of the morphogens (say, $X$) had to be autocatalytic and one (say, $Y$) had to inhibit the other. The autocatalytic morphogen $X$ reacted by converting chemical input $A$ into more $X$. Chemical inputs for this mechanism are crucial.

Even more amazing results might be possible if the collection of cells were immersed in a fluid. Would an elliptical sheet of cells ever develop a swimming instability (that is, a pattern in space and time that resembles the swimming of flatworms)? It seems likely, although the mathematical difficulty is very much greater than in the problem of the stationary ellipses describing *Drosophila* wing disks. The model of the flatworm might be an elastic elliptical sheet of muscle cells containing Turing's two morphogens, $X$ and $Y$, which control the twitching of contracting muscle cells. The sheet would be immersed in water and would behave according to hydrodynamic equations and to the reaction-diffusion equations of the morphogens. Morphogen $X$ triggers contractions of muscle cells when $X$ is sufficiently concentrated. The input concentration could be a food substance distributed in the medium outside the sheet. Then, could conditions be such that thermal fluctuations would trigger pattern formation in which the elliptical sheet "swims" through its food? This is the type of problem Turing's approach addresses, and the next generation of parallel processing in digital computers may make numerical simulation feasible.

These ideas extend the consequences of energy flow from the macromolecular level described in Chapters 2 and 3 to the tissue level of multicellular organisms.

## 4-3. DIMENSIONALITY AND INSTABILITY

A mathematical description of molecular evolution must take into account a very large number of variables (that is, it has many dimensions); but in some situations the dynamics is governed by only a few (1, 2, or 3) dominant variables (that is, it has few dimensions). These cases of small dimension are the best understood mathematically. Although the dimension of a simple model and that of a realistic model are vastly different, some features appear to be general. For example, self-organization of a morphogen pattern is conceptually related to the self-assembly of proteins. Both exhibit a tendency away from homogeneity toward asymmetric patterns, and both require an energy input. In self-assembly, the energy input results in protein synthesis; in self-organization, a chemical input results in an autocatalytic reaction step.

The consequences of changes in dimension illustrate some of the variations on general themes that dynamic systems exhibit. Turing used two morphogens because the generic reaction-diffusion behavior of a single morphogen concentration is stable uniformity (that is, the homogeneous, symmetric steady state (equilibrium) is asymptotically stable). To produce an instability requires at least two morphogen concentrations. This difference in behavior as a result of dimension (number of morphogen concentrations) is a provable mathematical theorem.

Dynamic systems of ordinary differential equations exhibit the effects of dimension explicitly. The one-dimensional case is

$$\dot{X} = f(X) \tag{3}$$

where $\dot{X}$ means $dX/dt$. This equation constitutes an initial value problem if $X \equiv X_0$ at $t = 0$. All bounded solutions to Equation (3) are monotone; that is, they asymptotically approach a fixed limit without oscillations (the second derivative with respect to time never changes sign).

The proof of the preceding claim follows as a special case of a general result in $n$ dimensions. In $n$ dimensions, a phase-space point has $n$ coordinates. Coupled first-order (having only first derivatives with respect to time) ordinary differential equations yield trajectories that never self-intersect. If the trajectory did intersect itself, then two different trajectories could use the intersection point as the initial value, which contradicts the existence of a unique solution. In one dimension, this implies monotonicity; that is, the trajectory cannot reverse its motion in one dimension for a first-order equation. In addition, many second-order systems (having second derivatives with respect to time) can be recast as higher-dimensional first-order systems.

In two dimensions the equations are

$$\begin{aligned} \dot{X} &= f(X, Y) \\ \dot{Y} &= g(X, Y) \end{aligned} \tag{4}$$

These equations permit oscillations, a behavior that reflects the related significance of two variables in Turing's theory.

Still other types of dynamic behavior are possible in higher dimensions. For example, trajectory bifurcation and the transition-to-chaos phenomena (discussed later) begin with three dimensions:

$$\dot{X} = f(X, Y, Z)$$
$$\dot{Y} = g(X, Y, Z) \quad (5)$$
$$\dot{Z} = h(X, Y, Z)$$

These types of behavior have biological importance that is discussed in Chapter 5.

### Mathematical Chaos

The complexity of trajectory orbits is characterized by numbers called *Liapunov exponents*. Given an initial phase-space point $(X_i, Y_i, Z_i)$, the rate at which a neighboring phase-space point $(X_i', Y_i', Z_i')$ separates from the initial point, as their two trajectories evolve in time, yields the Liapunov exponents $\sigma_i$, one for each dimension. The separation along the $i$th axis grows like $e^{\sigma_i t}$ (hence the name *exponent*). If $\sigma_i < 0$, then nearby points approach each other asymptotically in time; if $\sigma_i > 0$, then nearby points separate exponentially. *Chaos* is characterized as the occurrence of at least one Liapunov exponent greater than zero. Section 4-7 gives detailed examples of chaos.

Liapunov exponents also provide a measure of the rate of change of a volume of phase-space points as the volume moves through phase space according to some system of equations. The rate of volume change is denoted by $\Lambda_0$. If $\Lambda_0 > 0$, phase-space volume grows and becomes asymptotically unbounded. If $\Lambda_0 < 0$, then the evolution is a "contraction," which is typical of dissipative systems. The volume grows like $e^{\Lambda_0 t}$.

The chaotic states of interest are those occurring in dissipative systems. Thus, for at least one $i$, $\sigma_i > 0$, and $\Lambda_0 < 0$. The conventional numbering scheme for Liapunov exponents is to start with the largest, that is,

$$\sigma_1 > \sigma_2 > \sigma_3 > \cdots \quad (6)$$

Therefore, chaotic dissipative systems require $\sigma_1 > 0$ and $\Lambda_0 < 0$.

These quantities are connected by the identity

$$\Lambda_0 = \sum_{i=1}^{n} \sigma_i \quad (7)$$

because $e^{\Lambda_0 t} = e^{\sigma_1 t} \times e^{\sigma_2 t} \times e^{\sigma_3 t} \times \cdots \times e^{\sigma_n t}$. Using this identity to see an immediate consequence for dimension requires introducing one more technique of analysis.

Every system of $n$ coupled, first-order ordinary differential equations can be reinterpreted as a $n - 1$ dimensional Poincaré map. Bounded motions in

$n$ dimensions are of interest. A dissipative system, without inputs, simply decays to equilibrium. With inputs, it may have stable steady states, or it may have stable time-dependent states instead of stable homogeneous states. If its motions are to remain bounded, then some variables must be oscillating (generally nonharmonically) about some fixed value. The needed analytic technique is as follows: Observe the system, choose an oscillating variable, and choose a reference value for that variable in the range of its oscillations. Now, every time the value of the oscillating variable passes through the reference value, record the values of all the other variables. For example, let the variables be $X_1, X_2, \ldots, X_n$. Choose $X_1$ for observation, and choose the reference value $X_1^0$. Whenever $X_1(t) = X_1^0$, record $X_2(t), X_3(t), \ldots, X_n(t)$. At $t = 0$, $dX_1(0)/dt$ was either positive or negative. The record above contains points for which both signs of $dX_1/dt$ occur. Restricting the sign to that of $dX_1/dt$ at $t = 0$ results in a mapping of $(X_2, \ldots, X_n)$ that is the Poincaré map.

The prohibition on trajectory crossing discussed earlier does not preclude a common asymptotic equilibrium state. In fact, dissipative systems that are not driven do end up at a single-point equilibrium. All trajectories are disjoint except at their final, asymptotic limit, the common equilibrium point. An initial volume of $n$-dimensional phase space vanishes when all of it reaches the equilibrium state. In the driven case, the final state may be time dependent and may retain a finite "volume" (of lower dimension than $n$), a zero volume in dimension $n$ (more about this point later).

The Poincaré map also can be characterized by Liapunov exponents and by a volume $(2n - 1$ dimensional) contraction factor $\Lambda_0$. For the $n$-dimensional differential equation system, the $n - 1$ dimensional Poincaré map provides $n - 1$ Liapunov exponents, again denoted by $\sigma_1 > \sigma_2 > \cdots > \sigma_{n-1}$, although these are distinct from the $\sigma$'s for the $n$-dimensional differential flow. Again,

$$\Lambda_0 = \sum_{i=1}^{n-1} \sigma_i$$

A dissipative, chaotic Poincaré map must have $\Lambda_0 < 0$ and $\sigma_1 > 0$. The final step of the argument is that for a three-dimensional differential equation system, the two-dimensional Poincaré map must yield $\sigma_1 > 0$ and $\sigma_1 + \sigma_2 < 0$, which is a possibility; whereas for a two-dimensional differential equation system, the one-dimensional Poincaré map cannot yield both $\sigma_1 > 0$ and $\sigma_1 < 0$.

Therefore, a system that is characterized by only two degrees of freedom can exhibit patterned temporal behavior without the potential for chaos, whereas *all* higher-dimensional cases have the potential for chaotic driven states. Biological control mechanisms may have evolved to respect these mathematical differences in dimension: Restriction to two determinant degrees of freedom would produce control without risk of chaotic behavior.

## 4-4. TEMPORAL DEPENDENCE IN A DRIVEN SYSTEM: A TURING MODEL IN TIME

Consideration of simple two-dimensional dynamics for a system that remains spatially homogeneous focuses attention on the time dependence. The system is governed by the equations

$$\begin{aligned} x &= a + x^2 - yx \\ y &= xy - by \end{aligned} \tag{8}$$

The set includes an autocatalytic step, the $x^2$ term, and a driving term $a$. Both $a$ and $b$ are positive. Setting $\dot{x} = \dot{y} = 0$ and solving for $x$ and $y$ gives the steady-state values,

$$\begin{aligned} x^{ss} &= b \\ y^{ss} &= \frac{a + b^2}{b} \end{aligned} \tag{9}$$

Computing the linearized dynamics at $(x^{ss}, y^{ss})$ gives the stability matrix for this equilibrium state. The matrix is

$$\begin{pmatrix} b - a/b & -b \\ a/b + b & 0 \end{pmatrix} \tag{10}$$

and has the eigenvalues

$$\lambda_{\pm} = \tfrac{1}{2}\left(b - \frac{a}{b}\right) \pm \tfrac{1}{2}\left(-3b^2 - 6a + \frac{a^2}{b^2}\right)^{1/2} \tag{11}$$

which yields a pair of complex conjugate values.

When $a/b > b$, a perturbation near the equilibrium state will return to the equilibrium state by spiraling into it (Figure 4-1). The spiral motion is a combination of the negative real parts of $\lambda_{\pm}$ and the imaginary parts, which give rise to damped oscillations that look like spirals. For $b > a/b$, the real parts of $\lambda_{\pm}$ are positive, and the equilibrium is not stable. Instead, the dynamics produces an asymptotic time-dependent state. Although, in this case, the linearized solutions grow exponentially without limit, the true solutions to the nonlinear equations do not. When $a/b > b$, the spirals are counterclockwise in the $x$-$y$ plane and end up at the equilibrium point. When $b > a/b$, the trajectory also spirals away from $(b, (a + b^2)/b)$ in a counterclockwise sense, but the trajectory never settles down on a single point; it continues its counterclockwise bounded circuit forever.

When the dynamics approaches an equilibrium point in a stable way, the point is called a stable *fixed point* for the dynamics. When, instead, the dynamics

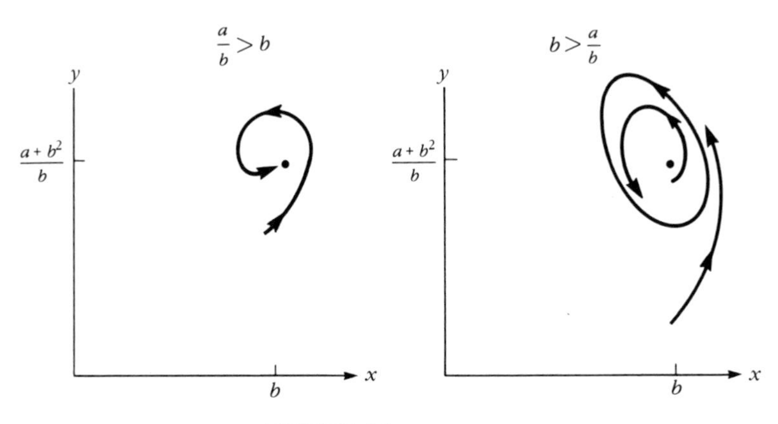


**FIGURE 4-1**
Fixed point and limit cycle.

forever produces an asymptotic time-dependent, closed trajectory, this trajectory is called a *limit cycle*. Figure 4-1 shows these cases for the present example. The limit cycle can be approached from inside or outside, depending upon the initial values for $x$ and $y$.

Much can be learned about this type of dynamic system even though there are no closed-form solutions.* The lack of closed-form solutions is not a result of insufficient cleverness but of their nonexistence. Only numerical solutions are possible, except in very special cases. In the preceding example, no closed-form solutions exist. However, graphical techniques and numerical computations on computers make the behavior of the system entirely accessible. A good qualitative understanding can be achieved with graphical methods alone, whereas quantitative understanding is aided by computers. For example, the particular system given in Equation (3) can be understood qualitatively by plotting the vector field $(\dot{x}, \dot{y})$ on the $x$-$y$ plane. Inspection of Equation (3) shows that the $x$-$y$ plane can be divided into four regions in which $\dot{x}$ and $\dot{y}$ each maintain the same sign (see Figure 4-2). The boundary lines at $x = b$ and $x = y$ are deduced from Equation (3). The figure shows that the motion around the fixed point $(b, (a + b^2)/b)$ is counterclockwise.

Another important feature of an unstable equilibrium state in a driven system is its behavior under thermal perturbations. Assume that the system starts in the state $(b, (a + b^2)/b)$ with $b > a/b$, then any thermal fluctuation that changes $(x, y)$ from $(b, (a + b^2)/b)$ will cause the state of the system to move away from $(b, (a + b^2)/b)$ and toward the limit cycle. The limit cycle is stable with respect to thermal fluctuations although it is somewhat smeared out by thermal noise.

Simulating the solution to Equation (3) on a microcomputer is easy. Plotting $x(t)$ and $y(t)$ against $t$ produces spiking for sufficiently large values of $b$ (Figure

* This is an analogue of the nonexistence of closed-form expressions for the roots of polynomials of degree 5 or higher, whereas such expressions do exist for degrees 2, 3, and 4.

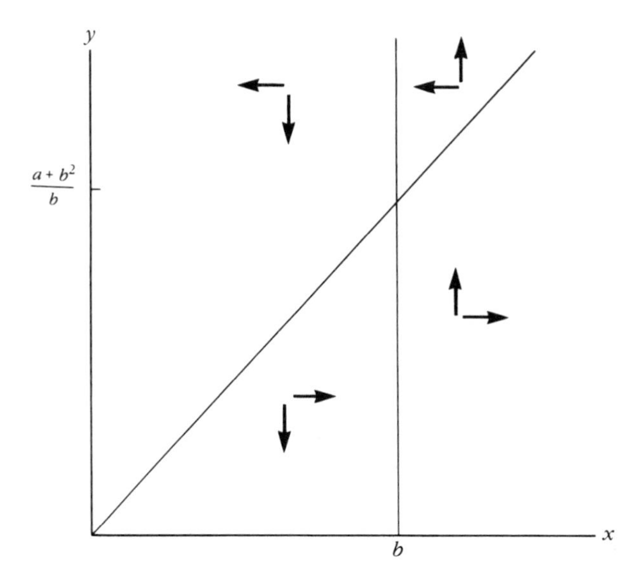


**FIGURE 4-2**
Qualitative graphical analysis. Vertical arrows indicate the sign of $\dot{y}$ and horizontal arrows the sign of $\dot{x}$; up is $+$ and down is $-$; right is $+$ and left is $-$.

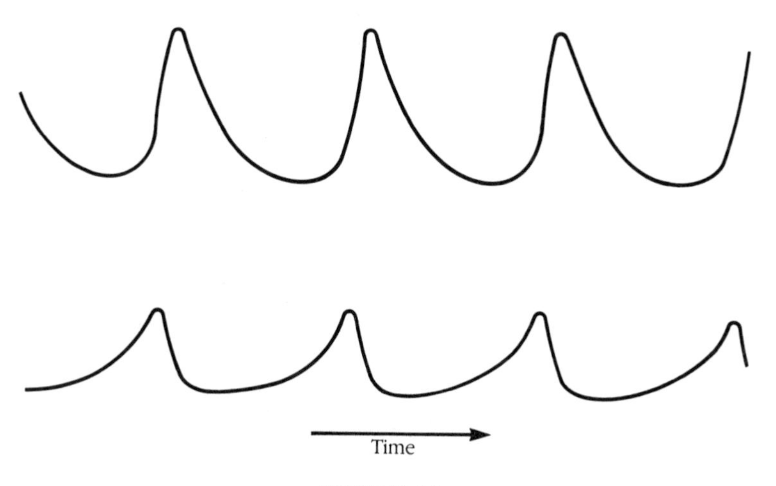


**FIGURE 4-3**
Spiking signals.

4-3). The time-dependent spiking pattern is a self-generated, emergent property of the nonlinear dynamics, even when all inputs are constants in time. Biological spiking patterns, such as in neurons, could also arise from such fundamental mathematical mechanisms.

## 4-5. ATTRACTORS IN PHASE-SPACE PICTURES

The geometric (graphical) approach to the theory of differential equations (especially ordinary differential equations) has led to the concept of an *attractor*: When a dynamic system has a stable fixed point, the attractor is that fixed point; and when the system has a stable limit cycle, the attractor is that limit cycle. For one- and two-dimensional systems of ordinary first-order differential equations, this exhausts the possibilities; but for higher dimensions, richer attractors exist.

In three dimensions, the so-called strange, or chaotic, attractor appears. A simple example of its behavior arises in the Rossler model (which has no physical or biological significance but is mathematically minimal),

$$\begin{aligned} \dot{x} &= -y - z \\ \dot{y} &= x + ay \\ \dot{z} &= b + z(x - c) \end{aligned} \tag{12}$$

in which $a$, $b$, and $c$ are positive constants. This extensively studied system shows interesting behavior for $a = b = 0.2$ and $c$ variable. Because this is a three-dimensional system, it is easiest to visualize its behavior on a two-dimensional plane; the $x$-$y$ plane is a natural choice. When $c = 2.6$, the trajectory projected onto the $x$-$y$ plane is a simple limit cycle. However, when $c = 3.6$, the projection is a double limit cycle. This doubling is called *bifurcation*. At $c = 4.1$, the projected trajectory has doubled again. The trajectories are shown in Figure 4-4, which also shows the asymptotic limit-cycle behavior (but not the transient approach to it for different initial conditions, which would clutter the figure). The self-crossing of the trajectory for $c = 3.6$ and $c = 4.1$ is not in conflict with the non-self-crossing requirement of first-order, ordinary differential equations because these are projected trajectories that do not cross in three-dimensional space.

While these trajectories are doubling in the $x$-$y$ plane, the behavior of the third variable $z$ is also changing. These changes are nicely described by the *power spectrum* of $z(t)$, which is found by taking the temporal Fourier transform of the correlation function for $z(t)$. The correlation function is $\langle z(t + s)z(t)\rangle$; this means to multiply $z(t)$ by $z(t + s)$ and then to average over all choices of $t$ for a given $s$, after transients (rapid initial changes) die out and the trajectory is on the limit cycle. The Fourier transform is then performed with respect to $s$. If $z(t)$ simply oscillates, then its power spectrum shows a single peak at the frequency of the oscillation. If, in addition, the equation for $z$ is

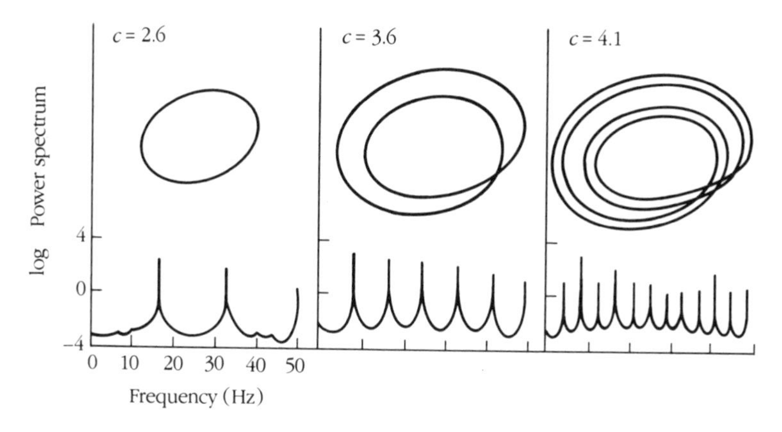


**FIGURE 4-4**
Bifurcations in the Rossler model.

nonlinear, then this nonlinearity can appear in the spectrum as the presence of higher harmonics of the base frequency. In the Rossler model, $z$ depends on the nonlinear term $zx$, whereas $x$ is linearly dependent on $z$. Therefore, if $z(t)$ has base frequency $\omega$, then so does $x(t)$, whereas the product $xz$ contributes a frequency of $2\omega$. This follows from simple trigonometric identities such as $\cos(\omega t)\cos(\omega t) = \frac{1}{2}(\cos(2\omega t) + 1)$. Because $z(t)$ contains the harmonic $2\omega$, so does $x$; and $xz$ contains $3\omega$ and $4\omega$ also. These higher-order harmonics are relatively weak compared with the base frequency in the power spectrum. All of this holds for $c = 2.6$ and is shown in the figure.

Something new happens when $c = 3.6$ or $c = 4.1$: Subharmonic frequencies appear in the power spectrum for $z(t)$. These correspond to the bifurcated trajectories in the $x$-$y$ plane. When $c = 2.6$, the trajectories are simple, and the time for tracing one cycle is $T = 2\pi/\omega$, where $\omega$ is the base frequency. For $c = 3.6$, the trajectory has bifurcated, and the time for tracing the entire trajectory is $2T$. This corresponds to a subharmonic frequency of $\omega/2$. For $c = 4.1$, tracing the entire doubly bifurcated trajectory, takes $4T$, which implies a frequency of $\omega/4$. However, nonlinearity, the $zx$ term, will generate superharmonics of these subharmonics, just as before, for $c = 2.6$. These superharmonics show up as frequencies such as $3\omega/4$, $3\omega/2$, and so on. The corresponding power spectra for $z(t)$ are shown in Figure 4-4, along with the $x$-$y$ trajectories.

This bifurcation sequence continues as $c$ increases beyond 4.1. In fact, bifurcation occurs more and more frequently for smaller and smaller increases in $c$. By $c = 4.6$, so many bifurcations have occurred that the $x$-$y$ plot looks like a smeared-out two-dimensional region of the $x$-$y$ plane. The corresponding power spectrum for $z(t)$ shows noisy contributions from an apparent continuum of frequencies (Figure 4-5).

The limit cycles in figure 4-4 look like one-dimensional curves in a two-dimensional plane. In figure 4-5, the attractor seems almost two dimensional. This apparent increase in attractor dimensionality is characteristic of attractors

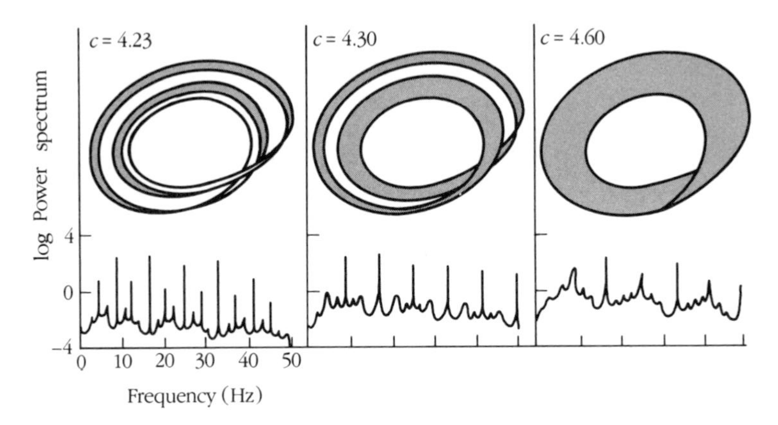


**FIGURE 4-5**
Chaos in the Rossler model.

that are the limit of an infinite sequence of bifurcated trajectories. Such an attractor is called strange or chaotic. Its dimension lies somewhere between 1 and 2; that is, it is a fractal dimension. Benoit Mandelbrot has discussed these fractal dimensions (Mandelbrot, 1982).

The chaotic attractor has still another curious property. Although transient trajectories simply move toward the attractor, either from outside or from inside, once they are on the attractor, they behave chaotically. This means that trajectories on the attractor that start very close to each other rapidly separate, although they remain bounded on the attractor. This separation of initially nearby trajectories is exponentially rapid during its initial phase, and the exponential rate determines a positive Liapunov exponent.

The dimension of an attractor is a mathematical concept. A real physical system, for which the equations are a model dynamics, always undergoes thermal fluctuations. Even when these fluctuations are small, they smear the trajectories. Consequently, for $c = 2.6$, a small amount of thermal noise causes the attracting limit cycle to have some width as well as length; the dimension, on average, appears to be 2 on the $x$-$y$ plot, not just 1. On the other hand, the smearing of the trajectories may be great enough so that the detail for $c = 4.23$ looks like the $x$-$y$ plot for $c = 4.30$ in figure 4-5, and the plot for $c = 4.1$ looks like the plot for $c = 4.23$. For a real physical or biological system, structures related to noisy attractors should appear, rather than those related to the purely mathematical attractor. However, the intrinsic noisy appearance of a chaotic attractor can be difficult to distinguish from the effect of extrinsic thermal noise. The intrinsic noise is solely a consequence of a driven nonlinear system.

A chaotic attractor is self-generating in the uroboros sense. Local regions diverge into other attractor regions because of chaos, while other regions arrive in the local region, regenerating it. The energy-driven state of the polymers envisaged in Chapter 2 was the attractor for the dynamics. In this case, the polymer not only provides the catalysts for the dynamics but also physically embodies the attractor.

## 4-6. CLOSED-FORM SOLUTIONS AND STRUCTURAL STABILITY

Nonlinear dynamic systems exhibit universality and special emergent properties simultaneously in their behaviors. For low dimensionality, these systems show strong dependence on dimension. On the other hand, given a particular dimension, all systems behave in classifiable ways, obeying known universality properties; this eliminates the need to study every possible $n$-dimensional equation, making it possible to use an extremely efficacious, topological criterion. In mathematics, this idea was originally promulgated by Henri Poincaré in his application of algebraic topology to the study of coupled first-order, ordinary differential equations.

The preceding ideas give substance to the mathematical concept of structural stability, which refers to the stability property of a differential equation under formal perturbations. For example, a damped oscillator is described by a pair of coupled first-order ordinary differential equations in which there is a damping parameter $\lambda$, as well as a mass $m$, a position $x$, a momentum $p$, and a frequency $\omega$:

$$\begin{aligned} \dot{x} &= \frac{p}{m} \\ \dot{p} &= -\lambda p - m\omega^2 x \end{aligned} \tag{13}$$

Finding the detailed behavior of this system for each value of $\lambda$ or $\omega$ is not necessary because all values of $\lambda > 0$ and $\omega \neq 0$ fall into two classes with equivalent topological behavior.* The so-called homotopy that exists between different pairs of values of $\lambda > 0$ and $\omega \neq 0$ in a class guarantees qualitatively identical behavior. This is one basic property of universality.

In one dimension, every first-order ordinary differential equation can be expressed as

$$\dot{x} = f(x) \tag{14}$$

where $f(x)$ is a single-valued smooth and continuous function of $x$. When $x = x_0$ at $t = 0$, the solution to Equation (14) constitutes the solution to the initial value problem in mathematics. The function $f(x)$, as given in Figure 4-6, has four local maxima and minima. (It may have none or many.) These extrema divide $f(x)$ into monotone segments. Over each such segment, a sim-

* The classes are $\lambda < 2\omega$ and $\lambda \geqslant 2\omega$.

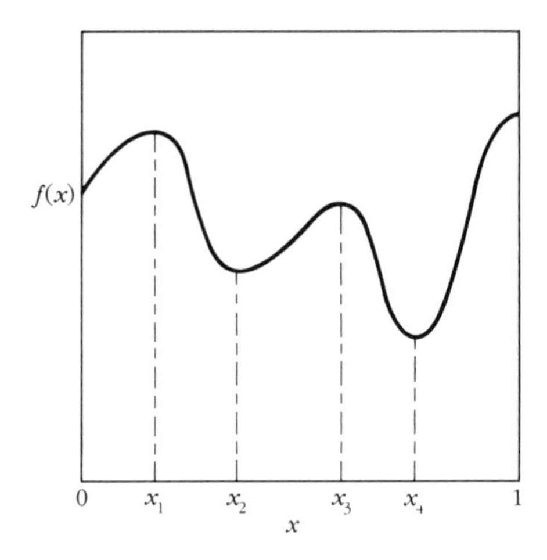


**FIGURE 4-6**
The function $f(x)$.

ple integral yields the solution

$$\int_{x_i}^{x_{i+1}} \frac{dx}{f(x)} = t_{i+1} - t_i \tag{15}$$

Thus, in one dimension, the problem completely reduces to the solution by integrals. This is the essence of universality, a single expression for all cases.

In two dimensions, universal reduction to the solution by integrals is lost. Many systems in two dimensions simply are not integrable or reducible to closed form. Numerical solution is still possible, but only within the bounds of computational accuracy. For example, the pendulum is an integrable, two-dimensional, hamiltonian dynamical system. Its momentum $p$ and angle variable $\phi$ satisfy the Newtonian equations,

$$\begin{aligned} \dot{p} &= -F \sin \phi \\ \dot{\phi} &= Gp \end{aligned} \tag{16}$$

These equations admit closed-form solutions given by elliptic integrals. The elliptic integrals generalize, to the nonlinear regime, the behavior of the trigonometric functions that describe a simple harmonic oscillator: Equation (16) approximated by $\sin \phi \rightarrow \phi$. Generalization results in an emergent behavior in the *separatrix* region of the phase-space picture of the pendulum system (Figure 4-7). In this picture, simple harmonic oscillations occur only in the region surrounding the origin; the closed ellipses show the trajectories for such motion. As the ellipses get larger, just inside the separatrix trajectories, the frequency of motion decreases greatly. At the separatrix, the period grows to infinity. Outside of the separatrix, the motion becomes a continuous rotation. The integrability implies the existence of a conserved quantity, in this

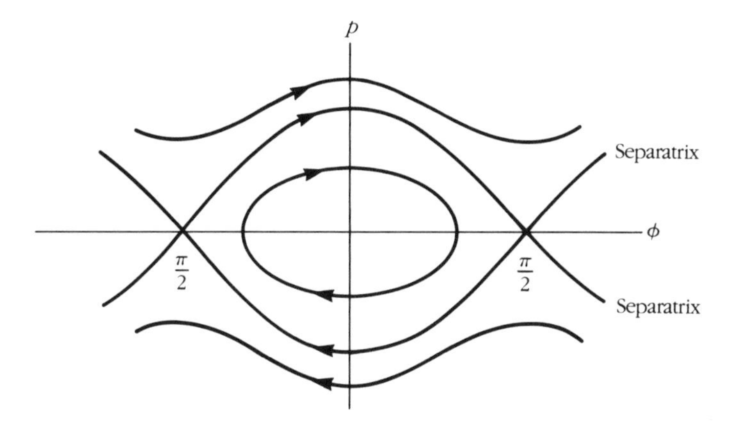


**FIGURE 4-7**
Pendulum phase-space portrait.

case the total energy, or hamiltonian, $H$:

$$H = \tfrac{1}{2}Gp^2 - F\cos\phi \tag{17}$$

This quantity remains constant in time. Using this identity to eliminate either $p$ or $\phi$ yields an equivalent one-dimensional problem, which is always integrable.

The damped pendulum, on the other hand, does not conserve total energy; it is a dissipative dynamic system described by the equations

$$\begin{aligned} \dot{p} &= -F\sin\phi - \lambda p \\ \dot{\phi} &= Gp \end{aligned} \tag{18}$$

in which $\lambda$ is the damping coefficient. There is a big difference between $\lambda = 0$ and $\lambda \neq 0$, which is the reason for excluding $\lambda = 0$ in the earlier discussion of the damped harmonic oscillator. However, in that case, the simple harmonic character of the $m^2x$ term in Equation (13) rendered the solutions reducible to closed form for both $\lambda > 0$ and $\lambda = 0$. In the damped pendulum, the two behaviors are qualitatively different. For $\lambda > 0$, the closed form is lost. Because a universal formula for the outcome in terms of the initial conditions and standard functions does not exist, numerical integrations are necessary. Nevertheless, all systems for which $\lambda > 0$ have identical topological behavior. In spite of the lack of a closed-form solution, the qualitative behavior of the dissipative system can be described; in this case, all initial states ultimately decay and end as damped oscillations around the origin.

In phase space, the difference between the integrable, hamiltonian case and the dissipative case is the difference between a cyclic orbit and a single point. Neither has any volume in the two-dimensional phase space, but the orbit is one dimension larger than a point.

Structural stability for a dynamic system, as a system of coupled ordinary differential equations, means that the qualitative behavior of the system is typical for an entire class of differential equations. The hamiltonian system is sensitive to dissipative perturbations and therefore is not structurally stable, whereas the dissipative dynamics is structurally stable under such perturbations. This is most easily seen if the dissipative system is energy driven. Structural stability is important in the context of this book because real physical systems are always subject to thermal perturbations, which makes them dissipative systems. The behavior expected in an energy-driven, thermally buffered physical system (that is, a driven dissipative system) is the behavior exhibited by the attractors of stochastic coupled differential equations. Thermal noise does not qualitatively modify the behavior of structurally stable systems of equations—just because they are structurally stable. The attractor structure, however, becomes smeared. A fixed point smears into a spot and a two-dimensional limit-cycle orbit smears into an annulus. With both, attractor dimensionality increases. In some sense, thermal noise affects the strange attractors of structurally stable dynamic systems by increasing the dimension of their fractal structure. They have a more noticeable volume in the full phase space.

Structural stability of the equations for dissipative systems implies that the nonexistence of closed-form solutions for a special case is no accident but is to be expected generally.

## 4-7. THE TRANSITION TO CHAOS: EXPERIMENTAL CASES

The method of Poincaré sections makes possible the detailed analysis of the behavior of periodically driven systems that are not integrable. The prototype for such systems is the periodically driven pendulum, which is governed by the coupled nonlinear equations

$$\begin{aligned} \dot{p} &= -F \sin \phi + A \cos \omega t \\ \dot{\phi} &= Gp \end{aligned} \tag{19}$$

in which $A$ is the amplitude of the periodic driving term of frequency $\omega$. This system, though not integrable, can be fully analyzed by Poincaré sections in the $p$-$\phi$ plane.

Letting $A = 0$ gives the pendulum dynamics, which is an autonomous system in two variables, $p$ and $\phi$. Values for $p$ and $\phi$ at $t = 0$ uniquely determine their values at any later time $t \geq 0$, in this case by closed-form expressions (elliptic functions). If the values of $p$ and $\phi$ at times $t_0$, $t_1$, and $t_2$ are $(p_0, \phi_0)$, $(p_1, \phi_1)$, and $(p_2, \phi_2)$, respectively, then $(p_1, \phi_1)$ is uniquely determined at time $t_1$ by the initial condition $(p_0, \phi_0)$ at time $t_0$. Moreover, with $(p_1, \phi_1)$

as the initial condition, then $(p_2, \phi_2)$ is uniquely determined at time $t_2$. This behavior means that the time evolution possesses the group property. (The group property means that time evolution over long times can be composed of products of time evolutions over short time segments.)

However, when $A \neq 0$, the initial condition $(p_0, \phi_0)$ at time $t_0$ uniquely determines $(p_1, \phi_1)$ at time $t_1$; but if $(p_1, \phi_1)$ is an initial condition, the value at time $t_2$ for $(p_2, \phi_2)$ is not obtainable without first carefully adjusting the argument of $A \cos(\omega t)$. Simply applying Equation (19) at time $t_1$ will not work, except for the special case in which $t_1 - t_0 = n(2\pi/\omega)$ and $n$ is an integer. This means that the group property of the time evolution, which was continuous for $A = 0$, is discrete for $A \neq 0$. This circumstance immediately suggests the Poincaré section idea.

Evaluating $(p_1, \phi_1)$ at time $t_1$ from $(p_0, \phi_0)$ at time $t_0$ for $t_1 - t_0 = 2\pi/\omega$, the period of the driving term, gives the output $(p_1, \phi_1)$, which can be used as the initial datum for time evolution to time $t_2$, providing that $t_2 - t_1 = 2\pi/\omega$ as well. Thus the sequence of phase-space points $(p_k, \phi_k)$, where $(p_k, \phi_k)$ corresponds to the values of $(p, \phi)$ at time $t_k = k(2\pi/\omega)$ with $k$ an integer, constitutes the Poincaré mapping for this periodically driven system.

Originally, Poincaré mapping was introduced for a dynamic system in $n$ dimensions in such a way that the Poincaré section was $n - 1$ dimensional; but in this analysis, the dynamic system and the Poincaré map are both two dimensional. This difference exists because Equation (19) is nonautonomous. Enlarging the dimension of the latter dynamics to give a three-dimensional, autonomous system with two-dimensional Poincaré sections recovers the previous definition. Using the simple artifice of introducing two new variables, $P$ and $Q$, achieves this increase in dimension and gives the new dynamic system

$$\begin{aligned} \dot{p} &= -F \sin \phi + P \\ \dot{\phi} &= Gp \\ \dot{P} &= -\omega Q \\ \dot{Q} &= \omega P \end{aligned} \tag{20}$$

This is a four-dimensional, autonomous system of coupled equations. By proper choice of the initial condition for $P$ and $Q$, the last two equations in (20) yields $P = A \cos \omega t$, and the equivalence of Equations (20) and (19) becomes evident. In addition, $P^2 + Q^2$ is a conserved quantity. This new conserved quantity implies that the system of equations is really three dimensional after all, as desired. Recording the $p$-$\phi$ values everytime that $P$ and $Q$ repeat themselves gives the Poincaré map for time intervals of $2\pi/\omega$.

This trick is useful for any dynamic system of $n$ dimensions that is periodically driven. Even for this minimal system, the driven pendulum, the increase in dimension is enough to allow chaotic trajectories for appropriately chosen values of $F$, $G$, $A$, and $\omega$. The transition to chaotic behavior begins with the generation of subharmonics, that is, with orbit bifurcations, as in the Rossler model discussed earlier.

## Emergent Behavior

Driven systems exhibit emergent behavior. A nonlinear system with constant inputs can exhibit nonconstant asymptotic behavior, and a periodically driven system can exhibit time-dependent behavior that shows many more frequencies than the single driving frequency. Neither of these behaviors appears in linear systems.

However, nonlinearity alone does not produce emergent behavior. A bona fide input term—that is, an energy-driving term—is required. After reviewing the situation in a number of model systems such as the Lorenz model,

$$\begin{aligned} \dot{x} &= -\sigma x + \sigma y \\ \dot{y} &= -xz + rx - y \\ \dot{z} &= xy - bz \end{aligned} \tag{21}$$

in which $\sigma$, $r$, and $b$ are constants, Vladimir I. Arnold reported the following:

> It seems that all models in which hyperbolic attracting sets have so far been found contain terms of the type of a pump or negative viscosity, which are absent in the Navier-Stokes equations. (Arnold, 1983, p. 275)

In this sentence, hyperbolic attracting sets are the strange attractors on which chaotic orbits exist, pump and negative viscosity are the energy inputs, and Navier-Stokes equations are the hydrodynamic equations in the absence of energy inputs. In fact, the Lorenz model, Equation (21), is the result of looking at the Navier-Stokes equations, with an energy input, and using only the slowest three modes of behavior.*

Definitive, necessary, and sufficient conditions for the transition to chaotic orbits are not known, however. Energy inputs appear to be necessary as well as nonlinearity. The transition, however, need not always follow period doubling (orbit bifurcation). In fact, several other transitions to chaos are known. For example, in an experimental study of hydromagnetic behavior in a periodically driven system, Albert Libchaber observed a different transition:

> Let us note again that for large Rayleigh numbers the states are always quasi-periodic, and that lock-in and period-doubling behavior is typical of small nonlinearities. (Libchaber et al., 1983, p. 82)

In this sentence, Rayleigh numbers measure the magnitude of energy input,

* A so-called Galerkin truncation.

quasiperiodic refers to time-dependent behavior with a Fourier spectrum made up of frequencies whose ratios are not rational numbers, and lock in refers to mode locking behavior (defined in Section 4-9). He also observed other transitions in the same system. In all, he reported four distinct types of behavior for this particular system.

Nevertheless, the list of transition types does not increase with each new system studied. Instead, only a few, perhaps as many as six, distinct types of transitional behavior exist. Sometimes a single system exhibits several types in various parameter regimes. Necessary and sufficient criteria for each type of behavior are not yet known, but the empirical approach has greatly narrowed the search, and a small number of distinct behaviors recur over and over in different systems.

Even the excursion into fractal dimension (see Section 4-5) is not purely mathematical. Real physical systems also exhibit those behaviors and structures. In fact, they exhibit all the curious behaviors mentioned here, including limit cycles, phase locking, bifurcations, and chaotic fractal attractors. This section concludes with examples of several physical systems, including a discussion of a variational principle approach.

***Example 1: Couette-Taylor Hydrodynamic System*** A fluid is contained between concentric cylinders that rotate independently (Figure 4-8). In most early experiments, the inner cylinder rotated while the outer cylinder remained fixed; that is, $\Omega_i > 0$ and $\Omega_o = 0$. Hydrodynamic flow between the cylinders is described by the Navier-Stokes nonlinear partial differential equations and suitable boundary conditions. One boundary condition is that a thin

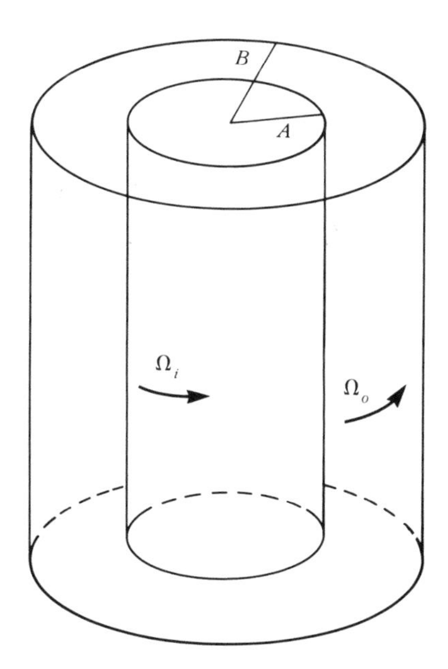


**FIGURE 4-8**
Couette-Taylor system. $\Omega_i$ and $\Omega_o$ are the inner and outer angular velocities, respectively, and $A$ and $B$ are the inner and outer radii, respectively.

layer of fluid "sticks" to the surface of each cylinder and has the same velocity as the cylinder. The lack of slip between the fluid layer and the cylinder creates a momentum shear in the fluid, which produces viscous heating (that is, energy dissipation). Thus work has to be done to keep a cylinder rotating. The machinery used includes servos to keep the rotation rate constant. In the analysis of the system, dimensionless quantities, the Reynolds numbers ($R_i$ and $R_o$ in this example) indicate when the flow will become unstable. These Reynolds numbers are

$$R_i = A(B - A)\frac{\Omega_i}{\nu}$$
$$R_o = B(B - A)\frac{\Omega_o}{\nu} \qquad (22)$$

in which $A$ is the radius of the inner cylinder and $\Omega_i$ its velocity, $B$ is the radius of the outer cylinder and $\Omega_o$ its velocity, and $\nu$ is the kinematic viscosity for the fluid in square centimeters per second ($cm^2/s$). In early experiments, $R_o = 0$ was often used, but richer behavior exists when $R_i \neq 0$ and $R_o \neq 0$. Moreover, there is no rotational relativity; that is, $R_i > 0$ and $R_o = 0$ are not equivalent to $R_i + \Delta R$ and $R_o = \Delta R$.

Nevertheless, this example focuses on the $R_o = 0$ case, in which the term *Reynolds number* means $R_i$, since $R_o = 0$. For small Reynolds number, a globally stable, unique solution exists to the Navier-Stokes equation and its stick boundary conditions. In cylindrical coordinates ($\rho$, $\theta$, $z$) the solution is

$$V_\rho = 0$$
$$V_z = 0 \qquad (23)$$
$$V_\theta = \frac{\Omega_i A^2 B^2}{B^2 - A^2}\frac{1}{r} - \frac{\Omega_i A^2}{B^2 - A^2} r$$

in which $r$ is the radius and lies between $A$ and $B$. This flow is called azimuthal laminar flow. As Reynolds number increases, and more work is done on the system, an instability occurs, precipitating a transition to Taylor vortices in the flow. These vortices resemble a stack of toroidal (doughnut-shaped) convective flows around the inner cylinder (see Figure 4-9). Alternate tori rotate in opposite directions. If the cylinders were infinitely tall, there would be infinitely many counterrotating stacked tori with identical radii and rotation frequencies. These features are determined by $R_i$, $A$, $B$, and $\nu$. The lowest value of $R$ for which this transition to vortices occurs is called the critical Reynolds number and is denoted by $R_c$.

In experimental work in 1965, Donald Coles (see Swinney, 1984) studied this hydrodynamic system in great detail. He found that for fixed $R > R_c$ many stable states occurred and that he could vary and control the number $N$ of vortices (finite for a finite cylinder). Furthermore, the vortices became unstable

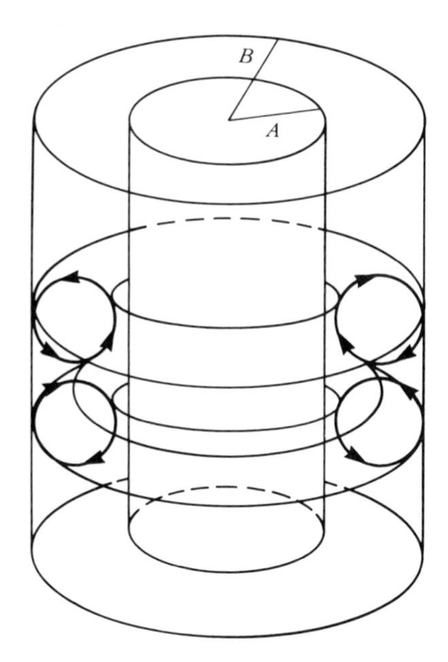


**FIGURE 4-9**
Taylor vortices in Couette flow.

to azimuthally propagating waves for sufficiently large $R$. These waves wrap around the inner cylinder axis an integral number ($m$) of times. Coles found a variety of interesting behavior in ($N$, $m$) space at fixed $R$.

Harry Swinney and others, using light-scattering equipment, have recorded spectra for these flows. In studies of correlations in the $V_z(t)$ projection of the velocity field, no signal occurs in the laminar flow case ($R < R_c$). Only a single frequency is present in the Taylor vortex case ($R > R_c$) unless the vortices are wavy, which means that a second frequency is present and that the nonlinearities have created integer mixtures of these two frequencies, such as $\omega_1 - \omega_2$ and $2\omega_2 - \omega_1$ (Figure 4-10).

The single-frequency case is periodic. The $\omega_1$, $\omega_2$ case is quasiperiodic. Lev Davidovich Landau proposed that the route to turbulence in strongly driven hydrodynamic systems occurred through the accumulated nonlinear mixing of successive frequencies. This view is called the Landau route to chaos and, symbolically, is $\omega_1$, $\omega_2$, $\omega_3$, $\omega_4$, . . . , chaos. For Couette-Taylor flow, if $R$ increases after the wavy-vortex regime sets in, chaos is usually sudden and full blown (that is, $\omega_1$, $\omega_2$, chaos). Just where $\omega_3$ would enter, the spectrum goes noisy. In Couette-Taylor flow, Gollub and Swinney (1975) were the first to clearly recognize this transition sequence, called the Ruelle-Takens route. The Ruelle-Takens scenario also allows $\omega_1$, $\omega_2$, $\omega_3$, chaos in some circumstances, but not $\omega_1$, $\omega_2$, $\omega_3$, $\omega_4$, chaos or higher-order analogues. Researchers have seen both of these cases, occasionally.

***Example 2: Rayleigh-Benard Convection*** Consider a horizontal layer of fluid heated from the underside. Heat conduction to the upper surface creates a thermal gradient. The lower and warmer part of the fluid becomes buoyant

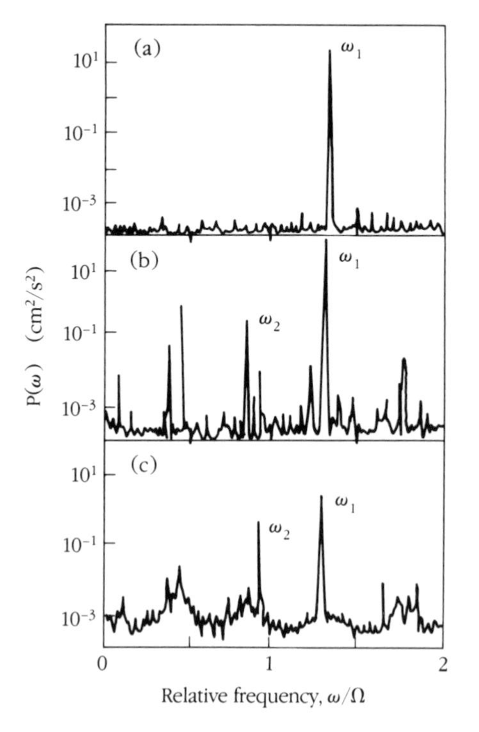


**FIGURE 4-10**
Taylor-Couette Power Spectra. Power spectra for various dynamic regimes in the Couette-Taylor systems. (a) $R/R_c = 9.6$, periodic; the spectrum consists of a single fundamental frequency $\omega_1$. (b) $R/R_c = 11.0$, quasiperiodic; the spectrum consists of two fundamental frequencies, $\omega_1$ and $\omega_2$, and integer combinations. (c) $R/R_c = 18.9$, chaotic; the spectrum contains broadband noise in addition to the sharp components $\omega_1$ and $\omega_2$. The noise in (a) and (b) is instrumental, whereas in (c) the fluid noise is well above the instrumental noise level. These spectra illustrate the transition sequence from periodic behavior through quasiperiodic to chaotic.

relative to the upper and colder part. For weak thermal gradients, in which temperature changes vary little with distance, only heat conduction occurs; the viscosity of the fluid prevents convection from taking place. For sufficiently strong thermal gradients, a convective instability develops, causing a transition into convective flow. An example of practical importance is the atmosphere. Consider the atmosphere to be a spinning, spherical layer that is periodically heated from below by radiation from the sun-warmed earth. Then the three-mode Lorenz model is the lowest-order approximation to the full dynamics.

Henri Benard performed the first experiments with Rayleigh-Benard convection in 1900 and Lord Rayleigh provided the correct interpretation, using the Navier-Stokes equations for this heat-driven system. In this case the dimensionless number that predicts stability is the *Rayleigh number*, which is given by

$$R = \frac{g\alpha\beta}{\kappa\nu} d^4 \tag{24}$$

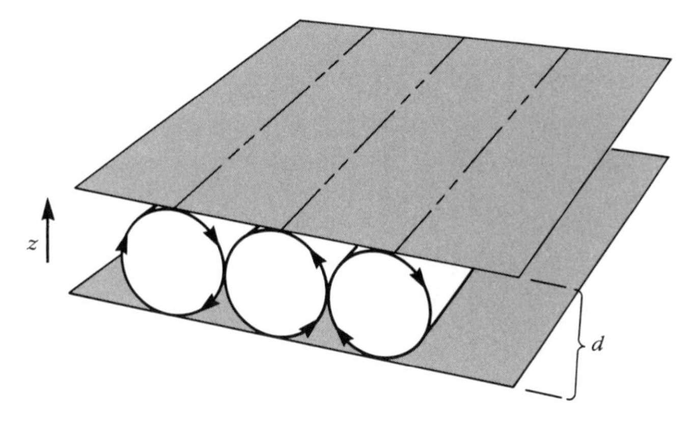


**FIGURE 4-11**
Benard rolls.

in which $g$ is the acceleration of gravity, $\alpha$ the volume expansivity, $\beta$ the temperature gradient ($\beta = | dT/dz |$), $\kappa$ the thermal diffusivity, $\nu$ the viscous diffusivity, and $d$ the layer thickness. Another dimensionless number, the Prandtl number $P = \nu/\kappa$ plays a role. A variety of behavior is found for different choices of $P$. For $P > P^*$ where $P^*$ is a critical Prandtl number, stationary convection arises, usually through constant counterrotating rolls (Figure 4-11). For $P < P^*$, however, time-dependent rolls that have waves propagating down their axes can occur.

Studies of Rayleigh-Benard convection have included work with spermaceti, mercury, and liquid crystals. In the last two cases, magnetic and electric fields respectively have been applied. Subrahmanyan Chandrasekhar (1961) beautifully presented his theory for the hydromagnetic behavior of mercury. Subsequently, Albert Libchaber et al. (1983) made another analysis that incorporated the more recently developed ideas about transitions to chaos; they observed period doubling, inverse bifurcations, Feigenbaum universality (Section 4-8), and mode softening (an additional route to chaos). Their analysis involves another dimensionless number, the Chandrasekhar number $Q$, given by

$$Q = \frac{\sigma B_0^2 d^2}{\rho \nu} \tag{25}$$

in which $\rho$ is the mass density, $B_0$ the magnetic field amplitude, and $\sigma$ the electrical conductivity. Low $Q$ values result in generation of subharmonics by bifurcation (period-doubling) and in mode-locking intervals. Higher values of $Q$ yield quasiperiodic behavior without mode locking. Figure 4-12 shows this behavior and the Ruelle-Takens route, in the form $\omega_1$, $\omega_2$, $\omega_3$, chaos. At low frequency, mode softening occurs, whereas at high $Q$ values quasiperiodic states go from $\omega_1$, $\omega_2$ to $\omega_1$, $\omega_2$, $\omega_3$, chaos.

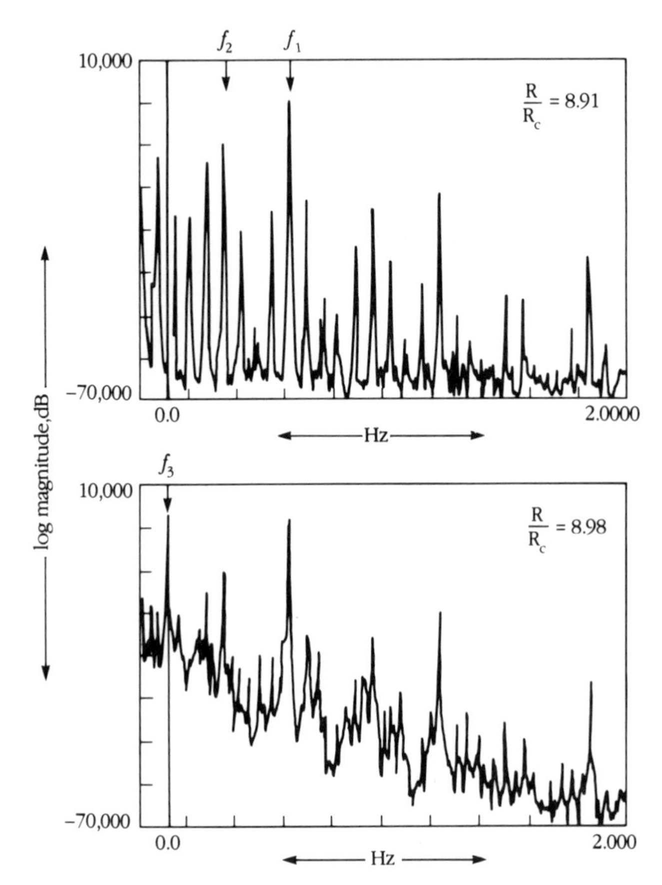


**FIGURE 4-12**
Quasiperiodic spectra for Rayleigh-Benard System. Fourier spectra corresponding to the Ruelle-Takens scenario. $R/R_c$ = 8.91 yields a quasiperiodic state with frequencies $f_1$ and $f_2$ (vertical line indicates position of future oscillator $f_3$); $R/R_c$ = 8.98, with oscillator $f_3$ present, yields exponential noise (shown by the constant slope in the recording).

Chandrasekhar also investigated the existence of a variational principle for characterizing the initial transition from conduction to convection in general Rayleigh-Benard systems. Consider a layer that has a weak thermal gradient across it. The layer is quiescent and conducts heat from the warm surface to the cooler surface. No net motion of matter occurs. The convective state, however, exhibits matter flow in counterrotating rolls. This matter flow is viscous (that is, dissipative) and damps out if the temperature gradient falls to zero. Heating from below drives the system by means of the buoyancy caused by gravity. The nonlinearity in the Navier-Stokes equations gives this system multiple stable solutions (stable states of convection). These convective states release free energy provided by the buoyancy force, which is larger than the viscous damping.

In these convective states, a matter flow pattern exists in which the buoyancy force is providing more energy than is dissipated viscously. How the transition from the quiescent to the convective state takes place poses an uroboros-type problem because the quiescent state cannot release the excess energy of buoyancy. Experimentally, increasing the thermal gradient produces the convective state. When the thermal gradient reaches the critical value $\beta_c$, any thermal fluctuation will kick the system from quiescent to convective. Chandrasekhar discovered the following principle from his variational approach:

> Thermal instability as stationary convection will set in at the minimum (adverse) temperature gradient which is necessary to maintain a balance between the rate of dissipation of energy by viscosity and the rate of liberation of the thermodynamically available energy by the buoyancy force acting on the fluid. (Chandrasekhar, 1961, p. 134)

As an example, assume that an experiment begins with no temperature gradient and with the layer in the quiescent state, which has no viscous damping. Then the researcher increases the temperature gradient, which eventually will pass through $\beta_c$. How will the fluid "know" when $\beta_c$ is reached, since $\beta_c$ involves the viscosity $\nu$? The answer is that the viscosity is continuously affecting the fluid at the microscopic level. Viscosity continuously damps momentum fluctuations in microscopic volumes of fluid, and thermal diffusivity $\kappa$ continuously damps heat fluctuations in microscopic volumes of fluid. That is, microscopic fluctuations at the critical thermal gradient cause the onset of convection.

The problem of initiating convection resembles the problem of initiating an autocatalytic process. In both, the energy released pays for the energy dissipated. It is uroboros-like because the first cycle of the stationary convective state must be initiated if the system is initially quiescent.

Discovering a comparable variational criterion for subsequent transitions between states that arise in a developing sequence would be desirable from the viewpoint of this book. Chandrasekhar has already made a few extensions of his approach, with some emphasis on energy-driven dissipative systems. But the last word certainly has not been written.

***Example 3: Belusov-Zhabotinski Reaction*** The Belusov-Zhabotinski reaction is a diffusing, autocatalytic oxidation of malonic acid by sodium bromate in an aqueous solution, with as many as 25 chemical intermediates forming along the way by means of at least 25 intermediate reactions. It requires only four reactants—ceric sulfate, sodium bromate, malonic acid, and sulfuric acid—and an indicator dye, ferroin, and is well worth observing firsthand.

A thin, unstirred layer of this solution in a petri dish exhibits propagating rings of alternating red and blue. These changes in the color of the indicator dye correspond to changes in the oxidation state of the fluid, which is diffusing and reacting on human spatial and temporal scales. Watching this reaction proceed is much more enlightening than still photography or systematic diagrams. The studies of Arthur Winfree and associates (Winfree and Strogate, 1984) uncovered beautiful structures, including scroll rings and toroidal vortices, in the propagating and reacting waves in three-dimensional media. The topological consideration of knots, twists, and links plays the central role in the analysis of these dynamic structures. Similar mathematical considerations have arisen in the study of supercoiling in double-stranded DNA.

Stirring the reactants to achieve spatial homogeneity reduces the problem to temporal evolution only, an approach used by Harry Swinney and his associates (Swinney, 1983). They used a flow-through reactor to control the flow of reactants through the stirred chamber and followed the state of the system in the chamber by measuring the bromide ion potential. In spite of the large number of intermediates and reactions involved, the bromide ion signal showed a regular behavior. At low flow rates, the potential oscillated at a single frequency. At higher rates, bifurcations appeared, followed by period doubling to chaos. The experimenters also observed mode locking (Section 4-9) and even the supercritical behavior of the Feigenbaum map (for $\lambda > \lambda^*$) (Section 4-8), the so-called $U$-sequence of $k$-cycles. Finally, Swinney and associates studied the attractors in the chaotic regimes for fractal dimension and for Liapunov exponents.

***Example 4: Cardiac Oscillator*** The excitation of cardiac tissue is modeled by a one-dimensional, two-parameter map,

$$\phi_{i+1} = \phi_i + b \sin(2\pi\phi_i) + \tau \tag{26}$$

which provides an analogue to the dynamic patterns observed in electrocardiograms of patients with dysrhythmias. These patterns often display period doubling and phase locking. (Electrocardiograms show the electrical potential measured at the surface of the skin above a local region of heart tissue.)

The work of Leon Glass and associates (Glass et al., 1983) has shown that the map, Equation (26), not only provides an analogue for the rich dynamics of dysrhythmias but also serves as a generic (in the topological sense) map for one-dimensional systems. As $b$ increases, the right-hand side of the equation eventually becomes noninvertible. The noninvertible regime is the most interesting.

The similarities between Equations (19) and (26) are not entirely accidental. Boris V. Chirikov (1979) has observed that the trigonometric nonlinearity, $\sin \phi$, is generic for "near-resonant" dynamics.

## Energy-Flow Ordering and the Transition to Chaos

In a driven system, a small amount of energy input leads to the spontaneous self-organization of space and space-time patterns. This is the meaning of energy-flow ordering. The ordered patterns result from energy-driven interactions in a dissipative substrate. If the inputs increase, transitions to new patterns occur; and with sufficiently large inputs, chaos is possible.

Closed-form mathematical solutions cannot describe even the nonlinear states that precede chaos. Understanding such states and, in particular, predicting their behaviors when no closed-form solutions exist are the subjects of the next sections. These issues will lead back to the biological significance of energy, especially of phosphagens.

## 4-8. FEIGENBAUM UNIVERSALITY

The discussion in Section 4-3 showed that an $n$-dimensional differential equation system gives rise to a $n - 1$-dimensional Poincaré map; in particular, a two-dimensional system produces a one-dimensional Poincaré map. Such a map cannot exhibit a transition to chaos because the dimensionality is too low.

But not all one-dimensional maps arise as Poincaré maps for two-dimensional differential equation systems. Many systems of much higher dimensionality possess an approximate behavior in which a single, slow variable satisfies the empirical map given by the difference equation,

$$x_{n+1} = f(x_n) \tag{27}$$

in which $f(x)$ is a noninvertible function with a graph like that in Figure 4-13. This map, called the logistic map, arose many years ago in the study of species population dynamics (May, 1976). The only requirements placed on $f(x)$ are that it be smooth and possess a single maximum with a nonvanishing second derivative at the maximum. The variable $x$ is normalized so that $f(x)$

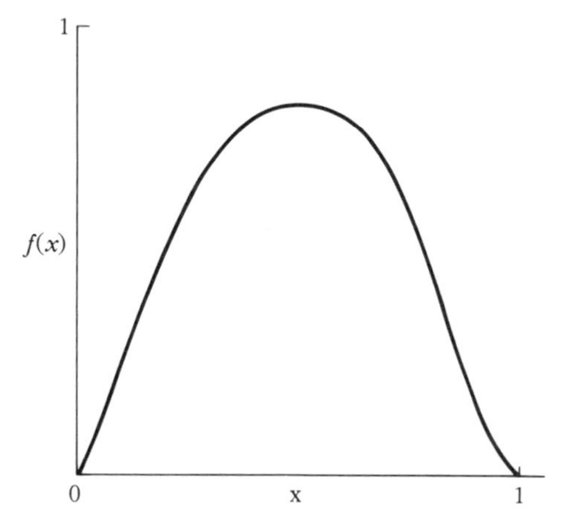


**FIGURE 4-13**
The logistic map.

$= 0$ at $x = 0$ and $x = 1$; $f(x)$ is normalized so that its maximum is bounded by 1. In general, $f(x)$ is also a function of a control parameter $\lambda$, which may take values in a specified domain. The simplest symmetric map of this type is the Feigenbaum map,

$$f(x) = 4\lambda x(1 - x) \tag{28}$$

in which $\lambda$ ranges from 0 to 1. The maximum for this particular map occurs at $x = \frac{1}{2}$, where $f(\frac{1}{2}) = \lambda$ and $f''(\frac{1}{2}) = 8\lambda \neq 0$ (unless $\lambda = 0$, a trivial case of no intrinsic interest). Using Equation (28) in Equation (27) gives the mapping,

$$x_{n+1} = 4\lambda x_n(1 - x_n) \tag{29}$$

### Computer Simulation

Remarkably, the preceding simple equation has no closed-form solution that is valid for all choices of $0 < \lambda \leq 1$, even though for each $\lambda$ in this domain and for any initial $x_0$ between 0 and 1, the sequence of values ($x_0, x_1, x_2, x_3, \ldots$) is uniquely determined by the equation. This sequence can be computed by hand, but a computer does it much faster. The program in BASIC is the following:

```
10  PRINT "WHAT IS L?"
20  INPUT L
30  PRINT "WHAT IS XØ"
40  INPUT XØ
50  X1 = 4*L*XØ*(1 -XØ)
60  PRINT X1
70  XØ = X1
80  GOTO 50
90  END
```

This program prints the sequence ($x_0, x_1, x_2, \ldots$) as far as you like. The reader may want to try it.

### Fixed Points

For $\lambda \leq 0.75 \equiv \frac{3}{4}$, any initial $x_0$ leads to a sequence that converges to a fixed point called $x^*$. The mapping thus has a fixed-point attractor. The value of $x^*$ is

$$x^* = 1 - \frac{1}{4\lambda} \tag{30}$$

which is easily verified. Equation (30) makes no sense for $\lambda < \frac{1}{4}$, because $x^*$ becomes negative; in fact, for $\lambda < \frac{1}{4}$, the attractor is actually $x^* = 0$. At $\lambda =$

$\frac{1}{4}$, $x^* = 0$ becomes unstable; so the attractor in Equation (30) is now stable. $x^* = 0$ is always a solution to Equation (29) but is stable only for $0 < \lambda < \frac{1}{4}$. Equation (30) is stable only for $\frac{1}{4} < \lambda < \frac{3}{4}$. The reader may figure out why this is so (it is quite easy) or may consult the original literature.

## Bifurcations

In the context of species population dynamics, $x$ is the population of a species, and the food source hides in $\lambda$. Therefore, $\lambda$ is the energy input parameter. As energy input increases, emergent behavior develops.

For $\frac{3}{4} < \lambda < 0.785$ something remarkable happens. The fixed point in Equation (30) becomes unstable; so no fixed point exists. However, asymptotically, any initial $x_0$ leads to an alternating pair of values $x_1^*$ and $x_2^*$ such that

$$\begin{aligned} x_2^* &= f(x_1^*) \\ x_1^* &= f(x_2^*) \end{aligned} \tag{31}$$

in which $f(x)$ is given by Equation (28). These are the fixed points of the iterate of $f$ with itself; that is, $f(f(x)) \equiv 4\lambda(4\lambda x(1 - x))(1 - 4\lambda x(1 - x))$. The mapping $f$ has no fixed points but bifurcates into a 2-cycle: $(x_1^*, x_2^*, x_1^*, x_2^*, \ldots)$. The value of $\lambda$ for which the 2-cycle first occurs is $\lambda_1$. At another value of $\lambda$, $\lambda_2 > \lambda_1$, the 2-cycle bifurcates into a 4-cycle. Both members of the 2-cycle, $x_1^*$ and $x_2^*$, bifurcate at the same value of $\lambda$. This also is easily proved. If $\lambda$ continues to increase, a sequence of $\lambda$'s results $(\lambda_1, \lambda_2, \lambda_3, \ldots)$, which marks the onset of successive bifurcations of each of the points of the cycle at the preceding stage. These $\lambda$'s produce 2-cycles, 4-cycles, 8-cycles, $\ldots$, $2^k$ cycles. The $\lambda_k$ get closer and closer together; thus infinitely many bifurcations occur before $\lambda$ reaches 1. In fact, this happens well below 1, at some $\lambda^*$.

## Universality

M. J. Feigenbaum (1983) discovered that if

$$\delta_n = \frac{\lambda_{n+1} - \lambda_n}{\lambda_{n+2} - \lambda_{n+1}} \tag{32}$$

then the limit $n \to \infty$ exists and

$$\delta = \lim_{n \to \infty} \delta_n = 4.6692016 \ldots \tag{33}$$

He also discovered that if a portion of a $2^{k+1}$ cycle is magnified and flipped (that is, rescaled), it nearly fits on top of the points of the $2^k$ cycle. As $k \to \infty$, this fit is increasingly better, and the rescaling parameter $\alpha$ has the value

$$\alpha = 2.502907875 \ldots \tag{34}$$

The amazing feature is that the Feigenbaum numbers $\delta$ and $\alpha$ are universal. They do not depend on the precise form of $f$ given by Equation (28). They occur for any one-hump map with nonvanishing second derivative at its maximum, such as in Figure 4-13, which Feigenbaum was able to prove.

What happens if $\lambda$ increases beyond the accumulation value $\lambda^*$? Mostly, chaotic behavior with no repeating cycles occurs. Scattered throughout this zone are $k$-cycles for all integer values of $k$, not just for values such as $2^k$ (for example, 3-cycles, 5-cycles, and 6-cycles); these constitute the so-called $U$-sequence. For example, $\lambda = 0.96$ in the Feigenbaum map, Equation (28), yields a 3-cycle. These new cycles also bifurcate repeatedly, although their $\lambda$ point sequences are much closer together initially than for sub-$\lambda^*$ bifurcations. Nevertheless, $\delta$ has the same value for each of them.

### Experimental Examples

All the features of universality have appeared in experiments with electrical circuits, chemical reactions, and hydrodynamic systems. Mathematical models, such as the Lorenz model, also exhibit this behavior. Experimenters have not observed much of the bifurcation sequence because thermal fluctuations smear out the $\lambda$ sequence, but they have obtained empirical measures of $\delta$, yielding, for example, $\delta = (R_8 - R_4)/(R_{16} - R_8) = 4.4 \pm 0.1$ for a hydromagnetic case in which $\lambda$ is replaced by $R$, a Rayleigh number (see Example 2 of Section 4-7).

Feigenbaum universality emphasizes several points: (1) A simple mechanism for sequences of bifurcations exists. (2) Feigenbaum universality occurs in many diverse mathematical and physical systems. (3) It does not require a precise mathematical form, but only a generic, broad class of forms. (4) Precise universal parameters characterize the bifurcation sequence. (5) No closed-form solutions exist in the chaotic regions. (6) Computer studies are of great utility, although graphical methods provide rich qualitative explanations of the behavior. (7) Asymptotically in time, the behavior approaches an attractor. (8) The attractor can be a fixed point, a $k$-cycle, or even chaotic (the limit of $2^k$-cycles as $k \to \infty$). Even so, still more types of behavior remain to be explored, as the next section shows.

Feigenbaum universality is also structurally stable (see Section 4-6). The effects of thermal noise can be incorporated as additive, inhomogeneous terms and as multiplicative (in $\lambda$) terms. Both cause smearing, and the latter shift in the $\lambda_i$ as well.

### Predictability

This example shows that even a simple mathematical mechanism can lead to apparently unpredictable behavior in the sense that no closed-form solution exists for the $\lambda$ values leading to chaos. To predict the population of a destructive species of insect year by year, when $x$ is the population (normalized to one) in a given year, the only approach, when the $\lambda$ value corresponds to a chaotic sequence, is to simulate the map on a computer. Of course, this simulation is much more rapid than real time; so for practical purposes prediction can be as accurate as desired. Thus *rapid simulation is the key to predicting the results of nonlinear and chaotic dynamical events*. Chapter 5 goes into the biological importance of this statement.

## 4-9. THE J. MAYNARD SMITH MODEL: INTEGER VERSION

The simplicity of the Feigenbaum map precludes certain types of behavior that occur only for higher dimensions. The J. Maynard Smith variation of the Feigenbaum map increases the dimension from one to two and introduces additional behavior. In the Feigenbaum map,

$$x_{n+1} = 4\lambda x_n(1 - x_n) \tag{35}$$

the factor $1 - x_n$ keeps the $x_j$ sequence bounded. J. Maynard Smith's variation is to introduce this inhibition term in a delayed way:

$$\begin{aligned} x_{n+1} &= 4\lambda x_n(1 - y_n) \\ y_{n+1} &= x_n \end{aligned} \tag{36}$$

The relative virtues of these maps in the context of species population dynamics are not important to this discussion; however, the species population dynamics language is amusing. In the Feigenbaum case, the reproductive inhibition is a function of the population of parents $x_n$, whereas in the J. Maynard Smith version, it is a function of the grandparents, because $y_n = x_{n-1}$.

Rewriting Equation (36) with slightly altered variables is convenient and useful for investigating its behavior: Replace $4\lambda$ by $\alpha$ and replace $(1 - y_n)$ by $(1 - y_n/N)$, in which $N$ is a fixed integer. The second change renormalizes the domains of $x$ and $y$ to a scale set by $N$ rather than by 1. Finally, Equation (36) becomes

$$\begin{aligned} x_{n+1} &= \mathrm{INT}\left[\alpha \frac{N - y_n}{N} x_n\right] \\ y_{n+1} &= x_n \end{aligned} \tag{37}$$

in which INT[. . .] means take the largest integer part of the argument. Equation (37) is a modified J. Maynard Smith model and maps the two-dimensional integer lattice onto itself. Discarding all real numbers except the integers reveals the fact that rich behavior does not require the continuum. The logistic map (Feigenbaum's map) is also expressible in integer form:

$$x_{n+1} = \mathrm{INT}\left[\alpha \frac{N - x_n}{N} x_n\right] \tag{38}$$

The reader should run this integer logistic map on a computer to observe how it bifurcates. Look, especially, at large $N$ ($\sim 10^6$ or more) as well as at $N = 1, 2, \ldots, 100, \ldots$. The integer nature of the map and lattice implies that an orbit has either a fixed point or a changing sequence of points. A changing sequence must repeat itself, eventually, because the phase space of the bounded integer lattice has only a finite number of points. Closed $k$-cycles must result, and these $k$-cycles occur with a spectra of rotation numbers, to be described below.

### The Computer Program

The BASIC program for Equation (37) is simply

```
 10 PRINT "WHAT IS A?"
 20 INPUT A
 30 PRINT "WHAT IS N?"
 40 INPUT N
 50 PRINT "WHAT IS XØ?"
 60 INPUT XØ
 70 PRINT "WHAT IS YØ?"
 80 INPUT YØ
 90 X1 = INT[A*(1-YØ/N)*XØ]
100 Y1 = XØ
110 PRINT X1,Y1
120 XØ = X1
130 YØ = Y1
140 GOTO 90
150 END
```

In place of the print statement in line 110, use a plot statement that plots X1 on the abscissa and Y1 on the ordinate. On my Apple II+ this statement is written

```
110 HPLOT 140 + X1, 120 - Y1        (39)
```

augmented with appropriate scale factors for N values greater than 100. Also needed is a line such as

```
85 HGR: HCOLOR=3        (40)
```

to go with 110. This routine can be as fancy as required, according to personal taste. The program to use in conjunction with the following discussion is

```
 5 HGR: HCOLOR=3
10 PRINT "WHAT IS N?"
20 INPUT N
```

```
 30  PRINT "WHAT IS A?"
 40  INPUT A
 50  PRINT "WHAT IS XØ?"
 60  INPUT XØ
 70  PRINT "WHAT IS YØ?"
 80  INPUT YØ
 85  I=Ø
 90  X1 = INT(A*(1-YØ/N)*XØ)
100  Y1 = XØ
110  HPLOT 100+2*XØ, 120-2*YØ
115  FOR W=1 TO 100
116  NEXT W
120  XØ = X1
130  YØ = Y1
135  I = I+1
136  IF I = INT(I/100)*100 THEN HGR
138  IF I = 199 THEN GOTO 150
140  GOTO 90
150  PRINT "N="N, "A="A: END
```

The graphics statement is now on line 5. I use an Apple II+ in its high-resolution graphics mode, HGR. It uses its finest line, HCOLOR=3, and plots XØ on the abscissa with YØ as the ordinate, as is shown by line 110; the constants provide centering and magnification. The new line 85 introduces an index for counting the number of times the program uses a computational loop. This index controls the graphics so that a clean, final, static image of the attractor results, from which the transient points of the trajectory have been subtracted. This is achieved by lines 135, 136, and 138. The HGR command in line 136 clears the screen after 100 iterations of the map. Because the microcomputer and its graphics display are so fast, I have been forced to throttle the process with a time delay. This is the purpose of lines 115 and 116. I have expanded the final line to make the program user-friendly.

## Representative Behavior

The behavior of this model depends upon every parameter: N, A, and the initial populations XØ and YØ. Begin by fixing N, XØ, and YØ (with N = 60, XØ = 2, and YØ = 2) and vary A.

For A less than 1, the initial point decays into the (0,0) state. At A = 1.1, the initial point is stable, and (2,2) is a fixed point. At A = 1.6, stability has moved to the point (21,21). The trajectory starting at (2,2) shoots up to the neighborhood of (21,21) and approaches on a decaying spiral trajectory. In a finite time, it reaches (21,21). Nearby initial points, for example, (2,3), perform qualitatively the same way and also end at (21,21); that is, (2,3) → (21,21). However, the initial point (2,30) immediately decays to (1,1). Remember that the $x$-coordinate is the autocatalytic variable, whereas the $y$-coordinate is inhibitory. This is similar to the dynamic structure of the Turing models. One must explore the initial value space for those regions that exhibit both bounded and interesting behavior.

### Emergence of a 6-Cycle

Let us continue with the initial state (2,2). At $A = 1.7$ a transition has already occurred, yielding a 6-cycle. The asymptotic fixed-point state has become unstable, producing a nonconstant 6-cycle, which cycles forever. This behavior persists with increasing $A$, although the size of the limit cycle—now a set of discrete points instead of continuous as in the Rossler model—increases as well. At $A = 2.0$ the 6-cycle has grown quite large, and the length of time (that is, the number of iterations required to eliminate the transient trajectory) becomes long. This is the analogue of critical slowing down in irreversible thermodynamics (the rate at which fluctuations decay at a critical point for a phase transition becomes very slow). To avoid artifacts, be careful not to use too small a denominator in line 136 (where I am using 100). For example, if the denominator is 40 and if in line 138 the 199 is replaced by 79, then one apparently gets a 14-cycle. The transient is not dead by the 40th iteration; but it is by the 100th iteration.

To make it possible to adjust this parameter rapidly when necessary, modify the program with

```
3   PRINT "WHAT IS M?"
4   INPUT M
136 IF I = INT(I/M)*M THEN HGR
138 IF I = 2*M-1 THEN GOTO 150
```

Choosing M = 100 gives the case being studied.

### Rotation Numbers

Table 4-1 contains the results recorded in terms of rotation number, which is defined as the ratio of the number of times the trajectory goes around its focus (the antecedent fixed point) to the total number of iterations required for one complete cycle. Thus the 6-cycle has rotation number $\frac{1}{6}$, and the stable fixed points have rotation number $\frac{1}{1}$. The interval from $A = 2.65$ to 2.2 has rotation number $\frac{1}{7}$, and the interval from 2.23 to 2.27 has rotation number $\frac{1}{8}$. Rotation numbers $\frac{1}{9}$ and $\frac{1}{10}$ also occur. Each of these cases occurs for a continuous range of $A$ values; so even though $A$ changes, the rotation number may stay the same. This is called *phase locking*, or *mode locking*.

Some phase-locking rotation numbers have numerators greater than unity. The first of these in Table 4-1 occurs for $A = 2.04$, where the rotation number is $\frac{4}{25}$; this corresponds to a 25-cycle that encircles its focus four times. Watching this event on the screen is worthwhile, and many other examples occur in the table. A rationale exists for the occurrence and values of rotation numbers.

Assume, for example, that $A = 2.11$. This value lies between $A = 2.0$ and $A = 2.18$, for which the rotation numbers are $\frac{1}{6}$ and $\frac{1}{7}$, respectively. For $A = 2.11$ the system is torn between 6-cycles and 7-cycles: It superimposes a 6-cycle on a 7-cycle by slightly altering them, yielding a 13-cycle with rotation number $\frac{2}{13}$. With $A = 2.08$, the trajectory would make two 6-cycles for each 7-cycle, which, when superimposed, would yield a 19-cycle with rotation number $\frac{3}{19}$, just as is observed for $A = 2.08$. The precise values for $A$ and the boundaries of the $A$ values for transitions in rotation number can also be determined numerically from the map given by Equation (37) and diophantine inequalities; however, this discussion will not delve into this arcane subject.

## Farey Addition

The occurrence of phase-locking intervals with rotation numbers possessing numerators larger than unity results from superposition of 6-cycles and 7-cycles, or 7-cycles and 8-cycles, or 8-cycles and 9-cycles, depending upon where in the $A$ domain the system is run. A simple arithmetic device exists for computing the expected rotation numbers: the Farey addition of rational numbers. Let two rational numbers, in lowest common factor form, be $p_1/q_1$ and $p_2/q_2$ where $p$ and $q$ are integers. Their Farey sum is

$$\frac{p_1}{q_1} \oplus \frac{p_2}{q_2} = \frac{p_1 + p_2}{q_1 + q_2} \tag{41}$$

Figure 4-14 lists the hierarchy of adjacent Farey sums; checks denote those found in Table 4-1. All can be found with sufficient effort.

The Farey sum table also omits other values for rotation numbers. At $A = 2.275$, the rotation number is $\frac{2}{18}$, which is quite distinct from $\frac{1}{9}$. The latter is a 9-cycle, and the former is an 18-cycle, which is a bifurcated $\frac{1}{9}$ case. For $N = 6{,}000{,}000$, a refined picture of the phase locking-intervals results; it is much closer to the Farey sum table but takes much longer to go through the evaluation procedure.

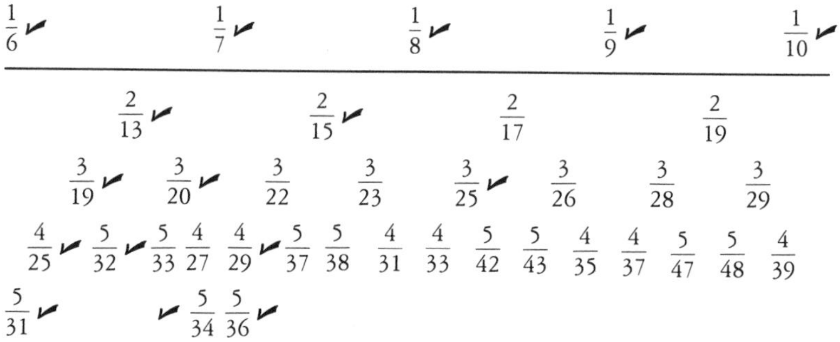

$\frac{1}{6}$ ✓ $\frac{1}{7}$ ✓ $\frac{1}{8}$ ✓ $\frac{1}{9}$ ✓ $\frac{1}{10}$ ✓

$\frac{2}{13}$ ✓ $\frac{2}{15}$ ✓ $\frac{2}{17}$ $\frac{2}{19}$

$\frac{3}{19}$ ✓ $\frac{3}{20}$ ✓ $\frac{3}{22}$ $\frac{3}{23}$ $\frac{3}{25}$ ✓ $\frac{3}{26}$ $\frac{3}{28}$ $\frac{3}{29}$

$\frac{4}{25}$ ✓ $\frac{5}{32}$ ✓ $\frac{5}{33}$ $\frac{4}{27}$ $\frac{4}{29}$ ✓ $\frac{5}{37}$ $\frac{5}{38}$ $\frac{4}{31}$ $\frac{4}{33}$ $\frac{5}{42}$ $\frac{5}{43}$ $\frac{4}{35}$ $\frac{4}{37}$ $\frac{5}{47}$ $\frac{5}{48}$ $\frac{4}{39}$

$\frac{5}{31}$ ✓ ✓ $\frac{5}{34}$ $\frac{5}{36}$ ✓

**FIGURE 4-14**
Farey summation of rotation numbers.

**TABLE 4-1**
***Rotation Numbers***

| *A* | *Rotation number* |
|---|---|
| 1.6 | $\frac{1}{1}$ |
| 1.63 | $\frac{1}{1}$ |
| 1.7 | $\frac{1}{6}$ |
| 1.8 | $\frac{1}{6}$ |
| 1.9 | $\frac{1}{6}$ |
| 2.0 | $\frac{1}{6}$ |
| 2.02 | $\frac{1}{6}$ |
| 2.03 | $\frac{1}{6}$ |
| 2.04 | $\frac{4}{25}$ |
| 2.045 | $\frac{6}{37}$ |
| 2.05 | $\frac{5}{31}$ |
| 2.06 | $\frac{4}{25}$ |
| 2.07 | $\frac{4}{25}$ |
| 2.075 | $\frac{3}{19}$ |
| 2.08 | $\frac{3}{19}$ |
| 2.09 | $\frac{3}{19}$ |
| 2.095 | $\frac{3}{19}$ |
| 2.1 | $\frac{5}{32}$ |
| 2.11 | $\frac{2}{13}$ |
| 2.115 | $\frac{2}{13}$ |
| 2.12 | $\frac{6}{39}$ |
| 2.125 | $\frac{3}{20}$ |
| 2.13 | $\frac{3}{20}$ |
| 2.14 | $\frac{3}{20}$ |
| 2.15 | $\frac{3}{20}$ |
| 2.155 | $\frac{5}{34}$ |
| 2.16 | $\frac{5}{34}$ |
| 2.165 | $\frac{1}{7}$ |

(*continued*)

**TABLE 4-1**
***Rotation Numbers (continued)***

| *A* | *Rotation number* |
|---|---|
| 2.17 | $\frac{7}{48}$ |
| 2.18 | $\frac{1}{7}$ |
| 2.19 | $\frac{1}{7}$ |
| 2.20 | $\frac{1}{7}$ |
| 2.21 | $\frac{5}{36}$ |
| 2.22 | $\frac{4}{29}$ |
| 2.23 | $\frac{1}{8}$ |
| 2.24 | $\frac{1}{8}$ |
| 2.25 | $\frac{1}{8}$ |
| 2.26 | $\frac{1}{8}$ |
| 2.27 | $\frac{1}{8}$ |
| 2.275 | $\frac{2}{18}$ |
| 2.28 | $\frac{3}{25}$ |
| 2.285 | $\frac{2}{18}$ |
| 2.29 | $\frac{2}{18}$ |
| 2.3 | $\frac{2}{18}$ |
| 2.305 | $\frac{1}{10}$ |
| 2.31 | $\frac{1}{9}$ |
| 2.32 | $\frac{1}{9}$ |

| *A* | *R.N.* |
|---|---|
| 2.212 | $\frac{1}{7}$ |
| 2.214 | $\frac{1}{7}$ |
| 2.216 | $\frac{4}{29}$ |
| 2.218 | $\frac{4}{29}$ |

| *A* | *R.N.* |
|---|---|
| 2.222 | $\frac{4}{29}$ |
| 2.223 | $\frac{2}{15}$ |
| 2.224 | $\frac{2}{15}$ |
| 2.225 | $\frac{2}{15}$ |
| 2.226 | $\frac{2}{15}$ |
| 2.227 | $\frac{2}{15}$ |
| 2.228 | $\frac{4}{30}$ |
| 2.229 | $\frac{4}{30}$ |
| 2.2295 | $\frac{4}{30}$ |

In the technical literature, the generic two-dimensional case (the two-dimensional analogue to Feigenbaum's map) is called the circle map. It is an analogue in the sense that it takes care of the two-dimensional case in general, just as Feigenbaum's map did for the one-dimensional case. Investigators have studied it extensively by computer simulation.

## Deterministic Stochasticity

To find arithmetic and number theory lurking in dynamic systems is amazing. Recall that this is the behavior of driven, dissipative systems which have structural stability and are classifiable by generic types at low dimension. Because such systems also have *maximal complexity* most of the time, their study requires computer simulation. Remarkably, the development of these ideas goes back to 1899 when Poincaré observed the complexity of behavior in dynamical systems possessing a homoclinic point. Such a point exists on the separatrix curve in the phase space picture of the driven pendulum, discussed earlier. In 1965, Stephen Smale published his account of the significance of Poincaré's observation (Smale, 1965). In essence he proved the existence of stochastic behavior (indistinguishable from a Bernoulli shift—that is, a coin toss) in certain deterministic systems. Such systems are now known to include driven, dissipative systems. Smale's discovery was that certain dynamic systems show a tendency to be torn between multiple fixed points by a stochastic decision mechanism. This, in effect, creates chaos in the solutions.

## 4-10. COMPLEXITY, CELLULAR AUTOMATA, DARWINIAN EVOLUTION, AND THE BRAIN

The J. Maynard Smith model in the integer version is extraordinarily simple: It takes two integers, combines them by a simple and explicit rule,

$$\begin{aligned} x_{n+1} &= \mathrm{INT}\left(\alpha \frac{N - y_n}{N} x_n\right) \\ y_{n+1} &= x_n \end{aligned} \tag{42}$$

and outputs two more integers. However, given $\alpha$, $N$, and initial values for $x_0$ and $y_0$, then knowing what to expect on, say, the 800th iteration requires calculating the entire trajectory. No general, universal (that is, closed-form) solutions are known that could yield the 800th iterative values (for example) from the input particulars without recourse to the preceding 799 values. This situation mimics the nonexistence of closed-form solutions in the case of continuous, differential equations. The next subsections show one way to take advantage of this feature.

### Complexity and Simulation

*Complexity* expresses the degree of difficulty in solving a problem, including the question of the existence of closed-form formulas into which data are put and the question of ease of using the formulas. Some explicit formulas are difficult to use in computation for a given degree of numerical accuracy. Because the J. Maynard Smith model has maximal complexity, simulating its solutions—inputting the data and iterating—is the best approach. However, in this case, a large number of iterations requires a very large amount of computing.

The word *simulate* is of key importance, since it implies control of the speed of the simulation. On a computer, the time per iteration may be so small that the 800th iteration is achieved in just a few seconds. Contemporary technology provides extremely fast simulation for two dimensions. But for higher dimensions, this approach is limited.

## Cellular Automata

Previously, we discussed the requirements of numerical integration for solutions to partial differential equations in space and time, such as reaction-diffusion equations (see Section 4-2). Operationally, this class of problems is the high-$n$, $n$-dimensional generalization of the two-dimensional problem in Equation (42). Many kinds of lattice problems have the same form. Examples include random walks in probability and statistical mechanics, and discrete space-time lattice-gauge theories for elementary particle physics. A number of computer-oriented scientists have already begun to catalog the types of behavior found for classes of lattice models. These classes serve the role of the homotopy classes in the topological classification scheme for differential equations and fall under the heading of cellular automata research.

The simplest of these models is essentially a one-dimensional Turing cellular model, that is, $n$ discrete sites in a line. At each site a finite number of variables exists, which can take on a continuum of values or, perhaps, a finite discrete set of values. In most of the early research, only one variable is present, and it has only two values. The investigator specified some rules for the dynamics of this system, which usually had the following form: (1) Specify each cellular variable value for the initial state; (2) determine new values for each variable by a system of coupled equations, such as in (42); and (3) use output values as initial data for the next iteration.

The researcher chose the equations, and much of this work was done as mathematics rather than as direct physical modeling. Often the studies were of nearest-neighbor interactions only, although some were of next-nearest-neighbor interactions or complex, multiply interconnected networks that were reminiscent of the complex wiring network of the brain. Recently, several reviews (Farmer et al., 1984) of this approach have appeared which provide further detail. The richness of the observed behavior of these simply-stated mechanisms is inspiring a rapid increase in interest in automata studies.

Once again, the richness can be classified and, surprisingly, the variety is not enormous, any more than was the case for the half dozen or so routes to chaos. Thus a search is underway for universality and for necessary and sufficient conditions. My own studies of these systems and their behavior have found that the observed evolutionary development of transient states is richest in driven, dissipative systems and that autocatalytic connections must be present, as well as engendered inhibitory connections. This is a recapitulation of the Turing mechanism for reaction-diffusion systems.

## Predictability for Nonlinear Dynamics

The concept of *predictability* concerns knowing the $n$th iterate before it occurs. Formulas provide instant predictability, whereas solutions by iteration do not. Most models suggested by physical and biological problems are maximally complex, so simulation is the only means available for predicting the outcome of events. The only requirement is for *rapid simulation*—a simulation that runs faster than the real-time system. Predicting the weather works in this way. A computer can simulate model equations much faster than the weather occurs, so prediction based on the model is possible. Indeed, this procedure is used for the national weather forecasts in the United States, although inadequacy of the models make long-term forecasting impossible.

### Predictability in Biology

The biological significance of the nervous system is that it can predict. Out of what necessity did it evolve? A need for simulation is attractive as the underlying cause. The muscular system aids an organism in coping with its environment and provides opportunities for destruction of an individual. A nervous-network mapping of each muscle, in conjunction with a nervous-network representation of a modeled environment, provide the substrate for simulation and concomitant muscular control. The simulation by the nervous system of muscle movements and their consequences in the environment must be faster than true muscle movements if effective control is to be achieved. Evolution has produced nervous systems for which this holds true; and mechanisms, both biochemical and physiological, make it so.

### Nervous System Coordination of Hydromedusae Movements

An evolutionarily early nervous system of relatively minimal structure was the ring of statocysts in the hydromedusae jelly-fish, which these creatures used to orient themselves relative to an axis defined by gravity. To sustain life they needed to know up from down and to make daily journeys between surface waters and deep waters. Perhaps this cycle was coupled to the sun, because some species also possessed a primitive light-sensing cell connected to the nervous ring.

The gravity coupling of the statocysts is also remarkable. A hollow cavity contains a slightly smaller, spherical ball (made from calcite, for example) which roles over nerve hairs fixed in the cavity. The hairs that are under the ball detect the orientation of the organism. The nervous ring is fast enough to coordinate a muscular response to environmental perturbations that disturb orientation. This simple nervous system coordinates motion without simulating potential outcomes.

### The Nervous System and the Transition from Arg-P to Cr-P

The nervous system of the worm is relatively complex compared with that of the hydromedusa (see Figure 5-1). Its network of interconnections is topologically richer than for a ring system. Recall (see Section 3-4) that the phosphagens show marked variations in the worm and that all higher forms of life use creatine phosphate (Cr-P) as phosphagen, instead of arginine phosphate (Arg-P) which is adequate for muscle alone. Did the transition to Cr-P lead to the ability of nervous tissue to simulate rapidly enough for sophisticated network function? To explore this question, consider the neuromuscular structure of the hydra, a simple coelenterate having a body made from just two layers of cells (Figure 4-15). The ectoderm and endoderm are not muscle tissue, per se, but contain abundant, organized muscle fibers, the action of which is controlled by a nerve net (Figure 4-16). From the simulation point of view, this nerve net looks like a lattice model for the muscular continuum. Its lattice structure is reminiscent of the cellular approach of Turing (Section 4-2) and of cellular automata (Section 4-10) models. From the phosphagen point of view, these organisms and the nerve net are Arg-P users; the need of this simple animal for rapid function is not too great. A more sophisticated organism might need more speed, and a more evolved phosphagen could be the key. Initially, a newly evolved phosphagen may have been important to the nervous system; but if it could make muscles react faster, it would also help the organism survive. Thus both tissue systems could benefit from using a substance like Cr-P. This is harmonious with the view that the evolution of phosphagens shows a direct correlation with the speed with which organisms can make use of them. In this instance, a direct connection may exist with the fact that Cr-P is the form in which phosphoryl groups are transported from mitochondria, the organelles of energy transduction, to the ADP in excitable components of cells. This transport property of Cr-P deserves emphasis.

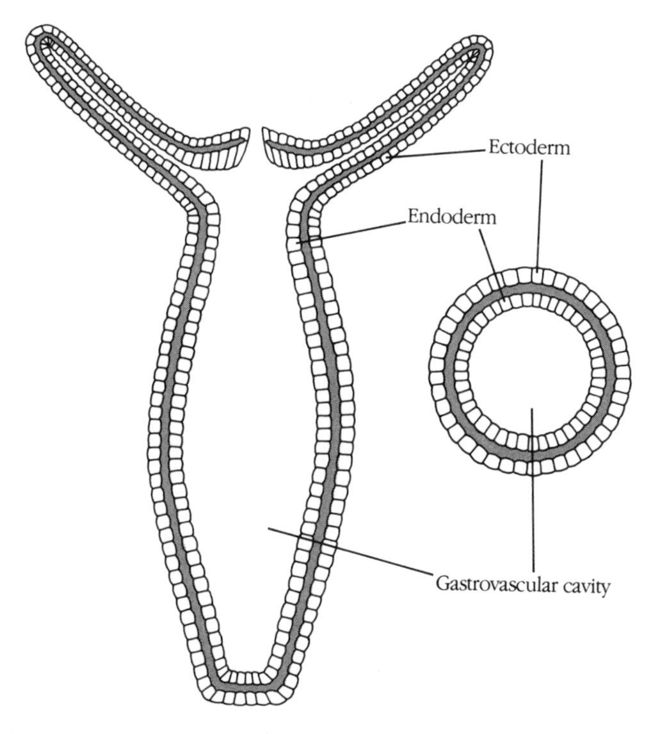


**FIGURE 4-15**
The hydra body consists of two layers of cells. *Left*, a hydra in longitudinal section; *right*, in cross-section. Between the two layers is a jellylike material.

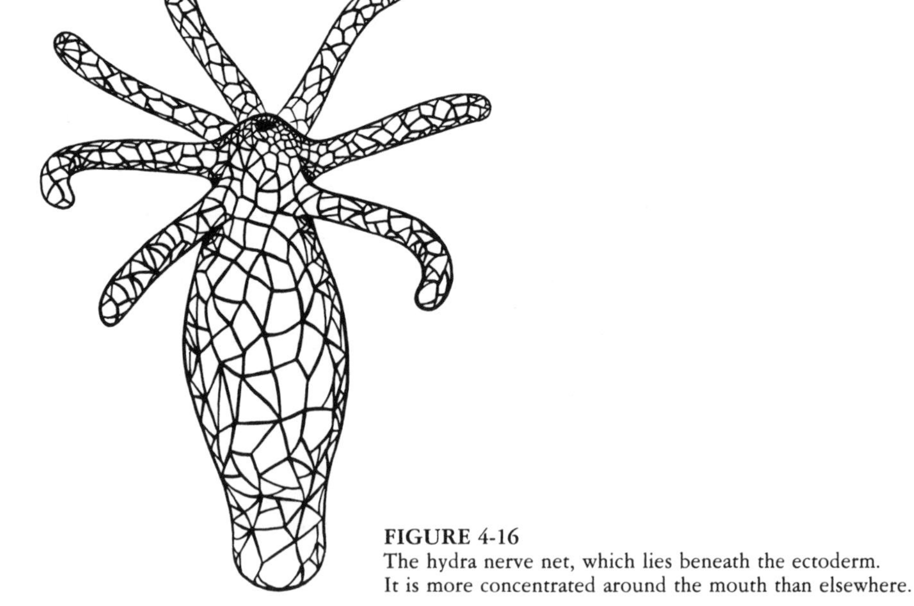

**FIGURE 4-16**
The hydra nerve net, which lies beneath the ectoderm. It is more concentrated around the mouth than elsewhere.

## Darwin's Struggle for Survival Reconsidered

Speed is not always the whole story in biology. Accuracy or fidelity is also important for survival. The criterion of Darwinian evolution, or survival of the fittest, would say that fitness requires both speed and accuracy. But ponderously slow, high-level fidelity would be no better or worse than rapid, nonspecific function as far as survival is concerned. Optimization with respect to these two requirements seems desirable but may not be as important as circumstantial events that occur in a diversified environment. Geophysical changes in the environment over geological time have often resulted in making a specific type of organism unfit whereas earlier it had been fit.

In Darwin's treatment of evolution, the emphasis on the struggle for existence is perhaps too strong. Ideas about the complexity of driven systems provide an alternative view. Energy-driven systems can evolve that exhibit elaborate sequences of events and diverse and variable structures. It is not surprising to see species come and go in a system as multidimensional as the surface of the earth and all its living organisms. This rise and decline of life forms is natural behavior for a driven system, although working out the interconnections among energy sources and the constraints may prove difficult.

In modeling, cellular automata can exhibit structures that appear, reproduce, and disappear as new structures appear. There is no struggle for existence. Although a simple, explicit set of rules predetermines the sequence of events, the variety of initial conditions available and the complexity of the resulting dynamics are such that no prediction of outcome is possible until a simulation of particular circumstances is run. Thus an uncertainty lies within this deterministic description. That is, determinism shows itself at each step of an iteration, but the eventual outcome appears only after a number of iterations.

The ability of the nervous tissue to simulate has provided the brain with novel Darwinian significance. In the strict Darwinian view, evolution proceeds by phenotype selection from stochastic variations in genotype. This mechanism describes very well the evolution at the unicellular level: the evolution of the metabolic pathways found in all organisms. But something new emerges in this system when nervous tissue evolves. The ability to simulate enables an organism to react to its environment in a fashion that transcends genetics because of the quicker time scale of simulation. Learning by an individual organism during its development clearly exemplifies this phenomenon in all advanced species. For example, apparently all the behavior of a hydra is wired-in genetically, but its ability to use the nerve net's simulation capacity to interact with the environment, although severely limited, does exhibit primitive learning (that is, conditioning). In more advanced forms, behavior is passed on generation to generation, as is shown, for example, by the visits of elephant herds to the Kitum cave salt licks in Africa, which have continued for thousands of years.

### Complex Behavior from Simple, Driven, Nonlinear Dynamic Systems

This chapter has explored the complex behavior of even very simple nonlinear dynamic systems; typically, there are no closed-form solutions to the governing equations, even though a relatively restricted variety of behaviors exists. The most important finding about the behavior of such systems is that predictability requires rapid simulation. I have proposed (1) that in biology the nervous system performs this function by simulating muscular activity and its consequences in the environment, (2) that this capability evolved with Cr-P, and (3) that it provides a new evolutionary mechanism that transcends Darwinian selection.

## REFERENCES

### Section 4-2

Kauffman, S. A., R. M. Shymko, and K. Trabert, "Control of Sequential Compartment Formation in Drosophila," *Science* 199 (1978): 259.

Thompson, D'Arcy, *On Growth and Form*, 2nd ed., Cambridge University Press, Cambridge, England, 1942.

Turing, A. M., "The Chemical Basis of Morphogenesis," *Philosophical Transactions of the Royal Society, London, Series B* 237 (1952): 37.

### Section 4-3

Lichtenberg, A. J., and M. A. Lieberman, *Regular and Stochastic Motion*, Springer-Verlag, New York, 1983.

### Section 4-5

Lichtenberg, A. J., and M. A. Lieberman, *Regular and Stochastic Motion*, Springer-Verlag, New York, 1983.

Mandelbrot, B. B., *The Fractal Geometry of Nature*, W. H. Freeman, New York, 1982.

### Section 4-6

Arnold, V. I., *Geometrical Methods in the Theory of Ordinary Differential Equations*, Springer-Verlag, New York, 1983.

### Section 4-7

Arnold, V. I., *Geometrical Methods in the Theory of Ordinary Differential Equations*, Springer-Verlag, New York, 1983.

Chandrasekhar, S., *Hydrodynamic and Hydromagnetic Stability*, Oxford University Press, London, 1961.

Chirikov, B. V., *Physics Reports* 52: (1979), 265.

Glass, L., M. Guevara, A. Shrier, and R. Perez, "Bifurcation and Chaos in Periodically Stimulated Cardiac Oscillator," *Physica D* 7: (1983), 89–101.

Gollub, J. P., and S. V. Benson, "Many Routes to Turbulent Convection," *Journal of Fluid Mechanics* 100: (1980), 449.

Gollub, J. P., and H. L. Swinney, "Onset of Turbulence in a Rotating Fluid," *Physical Review Letters* 35: (1975), 927.

Libchaber, A., S. Fauve, and C. Laroche, "Two-parameter Study of the Routes to Chaos," in *Order in Chaos*, edited by D. Campbell and H. Rose, North-Holland, Amsterdam, 1983.

Ruelle, D., and F. Takens, "On the Nature of Turbulence," *Communications in Mathematical Physics* 20: (1971), 167.

Swinney, H. L., "Observations of Order and Chaos in Nonlinear Systems," *Physica D* 7: (1983), 3–15.

Swinney, H. L., "Observations of Complex Dynamics and Chaos," in *Fundamental Problems in Statistical Mechanics VI*, edited by E. G. D. Cohen, North-Holland, Amsterdam, 1984.

Winfree, A. T., and S. H. Strogatz, "Organizing Centers for Three-Dimensional Chemical Waves," *Nature* 311: (1984), 611–615.

### Section 4-8

Feigenbaum, M. J., "Uiversal Behavior in Nonlinear Systems," in *Order in Chaos*, edited by D. Campbell and H. Rose, North-Holland, Amsterdam, 1983.

May, R. M., *Nature (London)* 261: (1976), 459.

### Section 4-9

Aronson, D. G., M. A. Chory, G. R. Hall, and R. P. McGhee, "Bifurcations from an Invariant Circle for Two-Parameter Families of Maps of the Plane: A Computer-Assisted Study," *Communications in Mathematical Physics* 83 (1982): 303–354.

Bohr, T., P. Bak, and M. H. Jensen, "Transition to Chaos by Interaction of Resonances in Dissipative Systems. I Circle Maps," *Physical Review A* 30 (1984): 1960–1969; and "Transition to Chaos by Interaction of Resonances in Dissipative Systems. II.

Josephson Junctions, Charge-Density Waves, and Standard Maps," *Physical Review A* 30 (1984): 1970–1981.
Guckenheimer, J., and P. Holmes, *Nonlinear Oscillations, Dynamical Systems, and Bifurcations of Vector Fields*, Springer-Verlag, New York, 1983.
Maynard Smith, J. *Mathematical Ideas in Biology,* Cambridge University Press, Cambridge, England, 1971.
Poincaré, H., *Les méthods nouvelle de la mécanique celeste*, Vol. 3, Gauthiers-Villars, Paris, 1899.
Smale, S., "Diffeomorphisms with Many Periodic Points. Differential and Combinatorial Topology," Princeton University Press, Princeton, N.J., 1965, pp. 63–80.

**Section 4-10**

Farmer, D., T. Toffoli, and S. Wolfram, eds., *Cellular Automata*, North-Holland, Amsterdam, 1984; *Physica D* 10 (1984).
Gardner, M., *Wheels, Life, and Other Mathematical Amusements*, W. H. Freeman, San Francisco, 1983.

CHAPTER

# 5

# *Biological Predictability*

Preceding chapters developed basic concepts and presented some of the evidence that supports the conclusion that a connection may exist between the mathematics of driven, dissipative systems and an understanding of biological evolution. This possible connection is examined again in this chapter.

The study of the origin of life, genetics, and protein synthesis led to considerations of energy metabolism and placed it at the heart of further conceptualization. Section 2-1 discussed the energized monomer to polymer transition and introduced ATP and ATP synthesis, which led to an appreciation of the evolved mechanisms of energy regulation and storage. These concepts imply the existence of an underlying physical impetus behind biological evolution. The physical impetus is energy flow in the form of phosphate bond energy, which manifests itself in a regulatory feedback network.

After the evolution of multicellular life, with its excitable tissues, the phosphoryl energy flow manifested itself through Cr-P. Simulation of nonlinear dynamic events, rather than polymerization, becomes the dynamic process of interest because motile multicellular organisms must cope with complex dynamic events. The mathematics of energy-driven, dissipative dynamic systems (Chapter 4) shows that the ability to predict events in such systems requires rapid simulation: Survival strategies include predicting possible results from hypothetical muscular actions. Mathematically, the only general approach is rapid simulation. I hypothesize that evolution is constrained to use this method and that this explains the evolution of the brain.

## 5-1. WHAT IS EVOLUTION?

*Evolution*, as the word is used by mathematicians and physicists is a process of change with time, but it has a more sharply defined meaning to the biochemically oriented biologist: It refers specifically to Darwinian selection, a genetic process whereby the off-spring of an organism show some variation of type that causes them to be selected, by favorable interaction with the environment, to survive and reproduce and give rise to new varieties.

### Environmental Evolution

The environment also evolves, in the physicists' sense, over geological time scales. Natural selection is a result of the interaction of organisms with their environments, and environmental evolution alters the nature of these interactions, thereby creating selection pressures that change with time. For example, plate tectonics has caused great environmental changes, over time scales of $10 \sim 100$ million years, resulting in part from the major climatic changes that attend the movements of plates through large differences in latitude.

### Genetics: The Mechanism of Evolution

Until the last half century, no one had correctly hypothesized any mechanism, but it has now become clear that the mechanism for Darwinian selection at the molecular level is to be found in molecular genetics. Genotype determines phenotype; variation in off-spring reflects variation in genotype, and the interaction of new variations with the environment determines the survivability of the mutants and thus the nature of later genotypes. Mutations can arise from a variety of causes, including random gene duplication errors, chemical influences, and cosmic rays. The basic notion is that variability is essentially random and the selection mechanism simply discards all those variations that do not work out as well as the original unvaried stock or as well as a few favorable variants. Consequently, the variants and nonvariants that reproduce most successfully dominate population dynamics.

Modern genetics has elucidated the molecular mechanism of evolution. Experimentors have artificially produced high rates of mutations in bacteria, fungi, and the fruit fly in laboratory environments designed to provide various selective pressures. These experiments clearly show selection among variants even though they have not run for long enough to derive new species through accumulated mutations. Nevertheless, the molecular biologist feels confident of the present detailed account of the molecular mechanism of evolution and has put this knowledge to use in medicine and agriculture. Its highest level of practical application is in genetic engineering.

## The Physical Impetus behind Evolution

The view of selection presented by Darwin was descriptive but not mechanistic at the level of genes. The recognition of genes and the proof of their significance came seventy years later, after which a rapid succession of insights took place: Investigators discovered the molecular structure of genes, explained their synthesis and function, and found proteins to be the agents and the products of gene expression. This led to an understanding of the role of energy in molecular biology.

Chapters 1 and 2 described how energy flow in an organic matrix could have led to the emergence of a primitive genetics and how energy flow of the proper kind could give rise to a molecular system that exhibits Darwinian selection as its mechanism of evolution.

From a mechanistic viewpoint, energy flow is the physical impetus behind biological evolution and sets the scene for the natural emergence of molecular control mechanisms. Chapter 3 discussed how these control mechanisms carry the signature of their energy-dependent origin. Their key component is phosphate, which is derived directly from energy metabolism. Phosphokinases are ubiquitous, and cyclic AMP (and cyclic GMP) figure prominently in a great variety of regulatory mechanisms, which researches now think have evolved from simple prototypes in prokaryotes into their eukaryotic manifestations, such as CAP protein function in operons.

Once biologists recognize that energy flow is the key to the existence of biological structures and their dynamics, it will not surprise them to see organisms evolving sophistications in their energy-processing techniques. Energy storage is especially interesting because it reflects, first of all, the fact that many organisms live and have lived in environments in which energy is and was abundant enough for energy storage to be feasible. The evolution of phosphagens, first in prokaryotes and then in eukaryotes, seems to be a key event in the emergence of multicellular life forms, because muscle function requires a phosphagen to supply energy rapidly. Muscle, an excitable tissue, gives an organism the ability to move rapidly only if the tissue can be properly energized, and this requires a phosphagen. Nervous tissue has a similar energy requirement. My view is that *energy storage, per se, was the antecedent to the evolution of excitable tissues*, and that Arg-P and Cr-P are the biochemical signatures of this evolution. In Chapter 3, I described how these phosphagens are responsible for the emergence of excitable tissues, which in turn naturally led to calcium-phosphate structures, a consequence of the energy role of phosphate, and the important role of calcium as an effector in energy mechanisms in excitable tissues. Thus the mechanism of Darwinian selection, applied to a system that exists only because of energy flow, has selected organisms having tissues that derived from progressively better methods of energy metabolism, use, control, and storage. This is the reason why evolution has taken life along the particular pathways it has.

I now argue that energy flow implies the existence of mechanisms for biological evolution that transcend Darwinian selection and its molecular genetics mechanism.

## 5-2. DRIVEN DISSIPATIVE SYSTEMS

Physically, biological evolution is a nonlinear, energy-driven, dissipative system. The dissipation in such systems is caused by thermal fluctuations; it plays a central role in determining reaction rates in aqueous milieu and causes the degradation of molecular structures (polymers), the existence of which is extremely unlikely on thermodynamic grounds in molecular mixtures at thermal equilibrium. However, biological systems are not in thermal equilibrium because they are energy driven. The structures and dynamic behavior of nonlinear, energy-driven, dissipative systems are utterly different from those typical of equilibrium systems made up from the same chemical elements. These ideas are embodied by the concept of the uroboros, a special kind of driven, nonlinear state.

Driven dissipative systems have properties that are determined by three simultaneous features: the constituent molecules (or chemical elements) of the system, the energy input, and the boundary conditions (environmental factors). If energy input and boundary conditions are fixed, then the dynamic system evolves in time toward an attractor. In the biological context, this attractor is noisy because of thermal fluctuations and is of very high dimension because biological systems are very complex. The high dimension reflects the great many degrees of freedom in a typical biological system, and the attractor that is approached may be a very complex state of matter. In reality, both the input and the boundary conditions also evolve, at least on geological time scales if not faster. In addition, the behavior of nonlinear, driven, dissipative systems exhibits emergent properties not predictable from the constituent molecules (or elements) by themselves in the absence of energy input. A most notable example is a system of biological polymers, which can only exist if precisely the right kind of energy flux (phosphate bonds) through monomers is supplied by energy transduction.

Emergent properties are characteristic of nonlinear, driven, dissipative systems. The great variety of macromolecular complexes and membranes and their self-assembly from their constituent metastable polymers dramatically illustrate that such properties exist in biology. Metastability is really very slow instability, and though proteins do hydrolyze, phosphate energy can drive their synthesis at rates much higher than that of hydrolysis. The assembly of metastable polymers obeys the second law of thermodynamics for closed systems as was discussed in Chapter 3 in some detail.

With energy input, self-organization can occur in self-assembled membranes or tissues. (Self-organization refers to processes of morphogenesis in a tissue or of dynamical activity in a membrane.) Turing showed chemical morphogens could cause a self-assembled tissue to exhibit waves that change in space and time. These waves self-organize within the self-assembled tissue, which would be homogeneously quiescent in the absence of the morphogen dynamics. The key to morphogen dynamics is energy input, which results in autocatalysis coupled to feedback inhibition. Section 4-4 showed how such mechanisms can cause constant-in-time input to result in nonharmonic, time-dependent dynamic behavior—a consequence that cannot happen in a linear system. Thus the emergent properties of nonlinear, driven, dissipative systems appear in at least three distinct ways: self-assembly, autocatalysis, and self-organization. Each manifestation is a consequence of energy flow; however, energy flow alone is not sufficient because the particular energy type and the particular substances involved determine the emergent properties.

Observing the richly varied dynamic behavior of nonlinear, driven, dissipative systems helps to understand biological evolution. That time-independent inputs can cause nonharmonic, time-dependent behavior in a nonlinear system is one of the most basic features of energy-driven systems. It may play a key role in morphogenesis (as proposed by Turing), in heart action, and in circadian rhythms.

## 5-3. THE EVOLUTIONARY IMPACT OF SIMULATION

The computer has greatly facilitated the study of nonlinear dynamic systems. This reflects the fact that no closed-form solutions exist for most nonlinear dynamic systems, which means that their behavior can be predicted quantitatively only if the dynamics is simulated. Simulation can use digital computing (conventional computers) or analogue models. In either case, the key to successful prediction of the behavior of nonlinear systems is *rapid* simulation—rapid enough that the outcome of simulation precedes in real time the outcome of the system studied. In biology, the need for prediction appears to have arisen in evolution only after simple multicellular organisms with muscle tissue had developed. Muscle tissue provides organisms with the motile ability to respond to environmental changes on a short time scale, such as in food gathering or in avoidance of being gathered for another organism's food.

The ability of muscle tissue to respond depends upon phosphagens. In the muscles of the simplest multicellular organisms, where the phosphagen role is played by Arg-P, rudimentary nervous systems control muscle tissue activity. Such nervous systems are also excitable tissues and also use Arg-P. Their function is, however, extremely rudimentary and can be characterized by just two types of neurons: sensory and motor. Sensory neurons are sensitive to gravity, light, touch, and chemicals, whereas motor neurons trigger excitation of muscle tissue. A simple organism such as a hydra or a jellyfish cannot respond in complex ways to situations; it is merely reflexive in its interaction with its environment. A more sophisticated organism—a worm, mollusk, or primitive fish—will be confronted with a much greater variety of environmental situations. Surely a more sophisticated nervous system would help the organism survive.

Chapter 3 followed the evolution of phosphagens to a stage that correlates strongly, both embyrologically and phylogenetically, with the advent of the chordates. The biochemical signature of this transition is the phosphagen Cr-P, and the structural concomitant of the energy-related nature of this development is bone (and cartilage) and a sophisticated nervous system.

## 5-4. NERVOUS SYSTEMS

In this section, I present the case that *the biological advantage of this advanced nervous system is to rapidly simulate the prediction of nonlinear events*. This step in the long history of biological evolution has critically altered the mechanism of evolution itself and transcends the genetic mechanism of Darwinian selection.

The mathematical studies of nonlinear, driven, dissipative systems have shown that changes in the value of an input parameter can cause sudden changes (transitions) in behavior: A static state can become time dependent, or a time-dependent state can become elaborate or even chaotic. In the biological realm, changes in energy transduction cause analogous changes in behavior. During evolution, the ability of organisms to transduce energy into biochemically useful energy (ATP) has improved so that the amount of input energy per cell per second has increased and is manifested by dramatic biodynamic changes.

An outstanding example of this increase in energy-input capability is energy storage by phosphagens. As I have argued (Section 3-4), the acquisition of Arg-P as a functioning phosphagen may have been a key step in the transition from unicellular to multicellular life forms. It is essential for the function of muscle tissue in times of high demand and is also found prominently associated with nervous tissue, even in the simplest invertebrates, including hydra, worm, and jellyfish. The hydra's nervous system was shown in Figure 4-15. Worms possess simple, segmented nervous systems (Figure 5-1), and some jellyfish posses only a ring of neurons.

The polychaetes and the octopus, which employ unusual phosphagens, show signs of a more sophisticated nervous system containing segmented, disjoint,

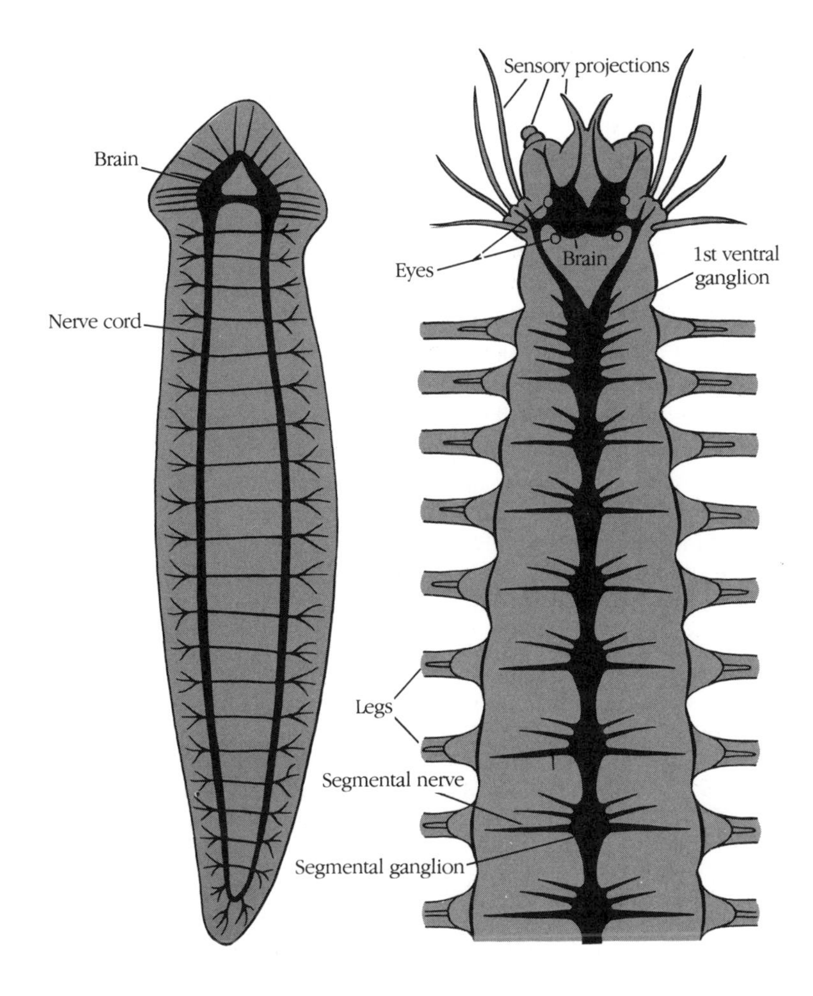


**FIGURE 5-1**
Primitive nervous systems. (*a*) Nervous system of a planaria. (*b*) Nervous system of a nereis.

and rudimentary ganglia. Here, too, the tissues in which phosphagens are abundantly found are the excitable tissues, muscle and brain.

The evolution of central nervous systems corresponds with the rise of the Chordata, all of which use Cr-P. Increasing the rate of energy flow by changing from Arg-P to Cr-P appears to have been the impetus for the transition from a primitive to a sophisticated neuromuscular system containing a central processing cortex.

## The Intermediate Neural Network

The chordate nervous system consists mostly of the intermediate neurons, which connect sensory neurons with motor neurons. Higher primates, for example, have around $5 \times 10^6$ motor neurons and $10^{10}$ intermediate neurons, or about 2000 intermediate neurons per motor neuron. The central nervous system is structurally similar throughout the Chordata phyla, although major differences in the sizes of various parts exist, particularly in the cerebrum; but overall the design has been conserved throughout evolution (Figure 5-2). (Note that the six examples in the figure do not constitute a sequence of lineal descendants; they are a collection of off-shoots from the main lineage and therefore are only suggestive of an evolutionary sequence.)

The organization of the chordate nervous system is based on a spinal cord connected to a central processor, the brain. The spinal cord, reminiscent of the segmented nervous system of worms, connects to the limbs and torso by sensory and motor nerves, which interconnect at the spinal column (Figure 5-3). The spinal cord connects to the brain through the brain stem, which attaches to the limbic system (rhombencephalon), off of which extend the cerebellum, the diencephelon, and the cerebral cortex. All these parts make up the brain, which is composed almost totally of intermediate neurons (Figure 5-3b). According to C. Judson Herrick (1922), the appearance of the "great intermediate net" marks a fundamental step in the evolution of the nervous system. The cerebellum contains layers of Purkinje cells (Figure 5-4) and is dedicated to the integration of motor control. The limbic system, which controls instinct and mood, also contains a small number of cell layers. The cerebrum is multilayered; the layered structure underlies its function as processor and simulator.

## The Simulator Architecture

In elucidating the process of motor control, neurobiologists identified many different types of intermediate neurons, their morphology, and the anatomy of their interconnections. The Purkinje cells play a central role in learning coordinated motor behavior and in determining congenitally fixed movements. Clear evidence shows that this part of the nervous system has produced a sensory afferent mapping of the organism's body in the brain and a motor efferent mapping from the brain back to the body (Figure 5-5). The afferent and efferent fibers connect at the spinal cord, and from there connections run into the brain (for instance, to the cerebellum). The cerebrum, as well as the limbic system, appears to provide several additional kinds of mapping of brain functions onto itself. The visual system, an extension of the brain, provides an image of the organism's environment and as such constitutes a mapping of the environment into the brain. Motor control uses interactions with the visual system, the cortex of which is multilayered (Figure 5-6), much like the rest of the cerebral cortex.

The coupling of the visual system and the motor system permits the brain to use its capacity for rapid simulation to anticipate the outcome of movements before they actually occur. Through cortical representations of organism-environment collisions, difficulties can be anticipated and avoided. Since the dynamics of organism movements and their interactions with the dynamics of

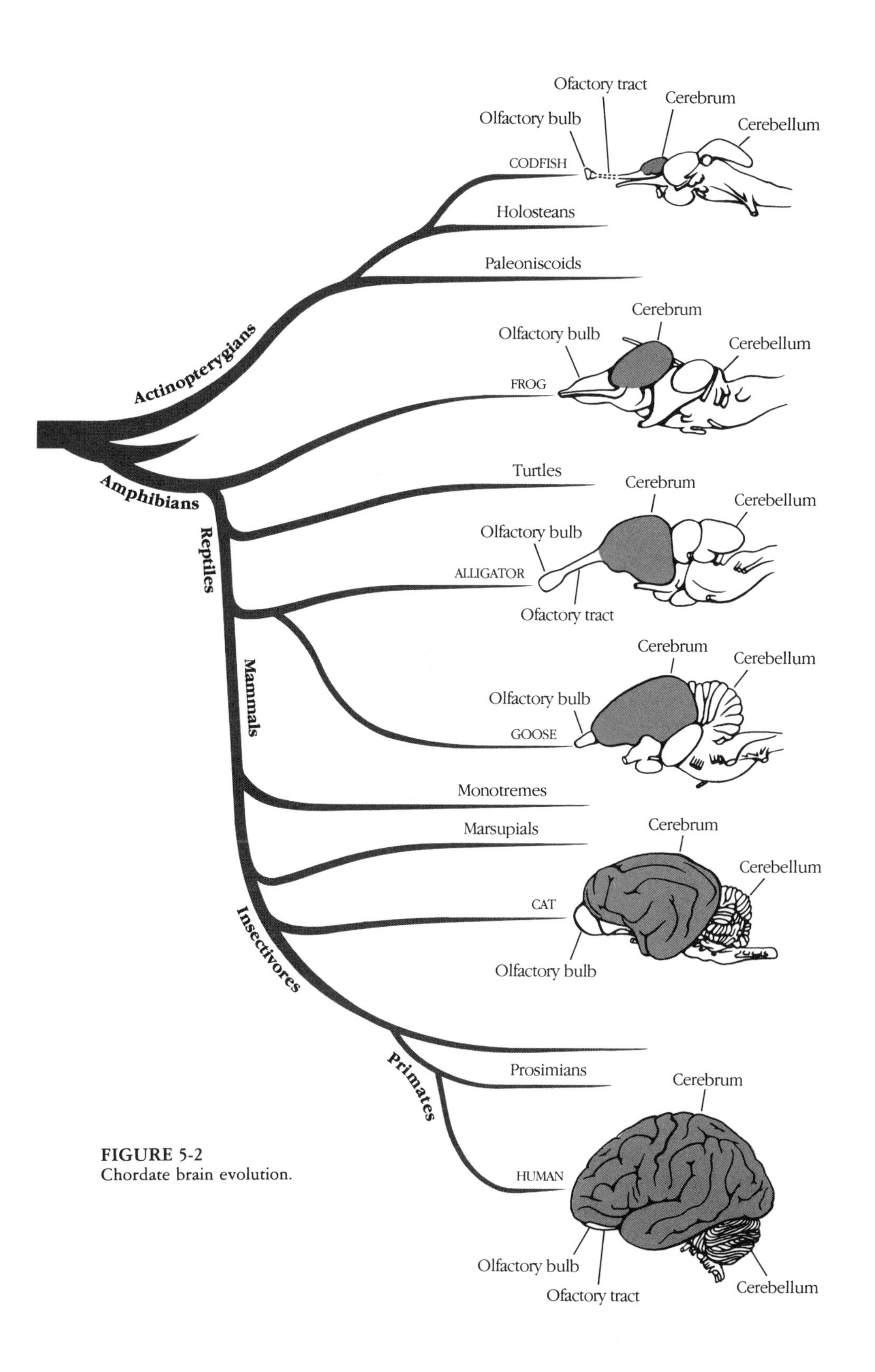


**FIGURE 5-2**
Chordate brain evolution.

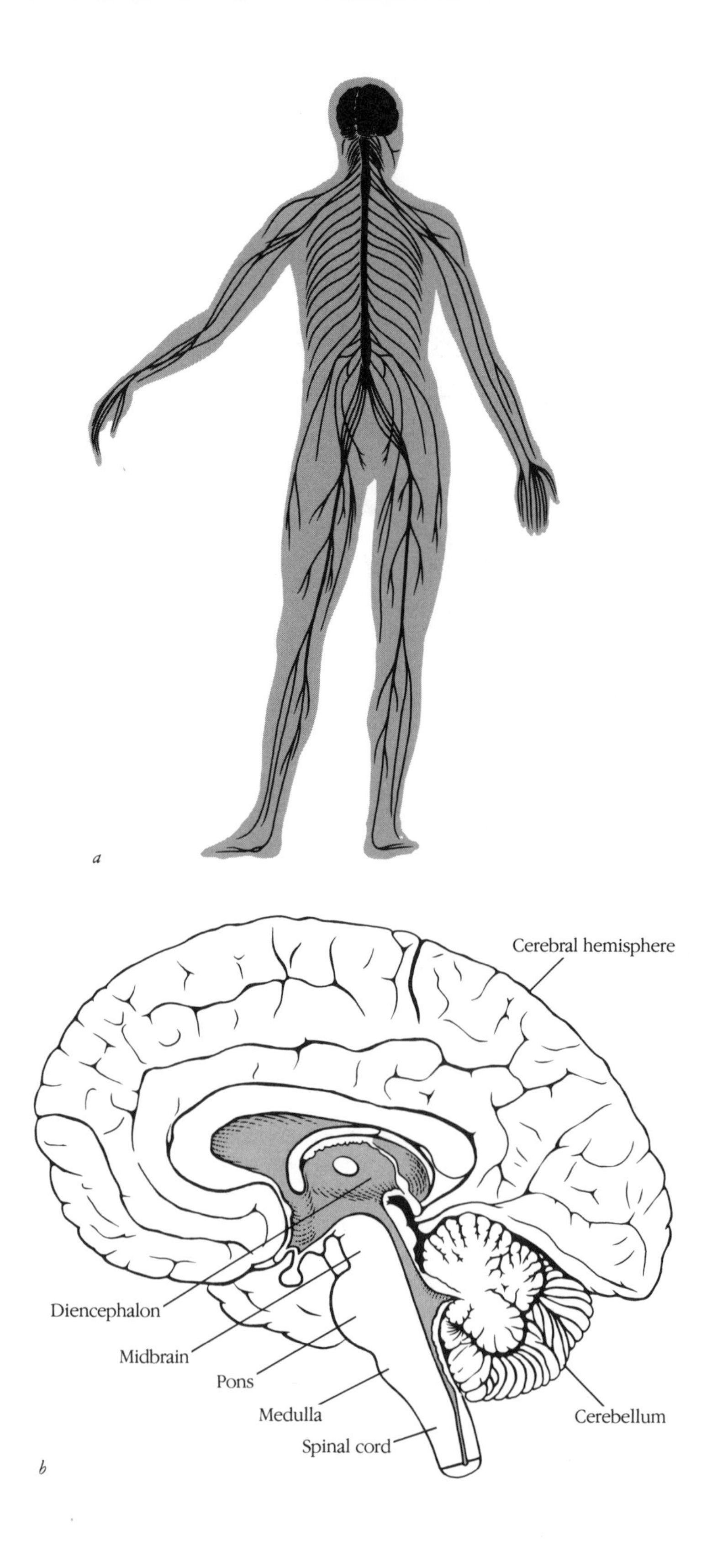
a
Cerebral hemisphere
Diencephalon
Midbrain
Pons
Medulla
Spinal cord
Cerebellum
b

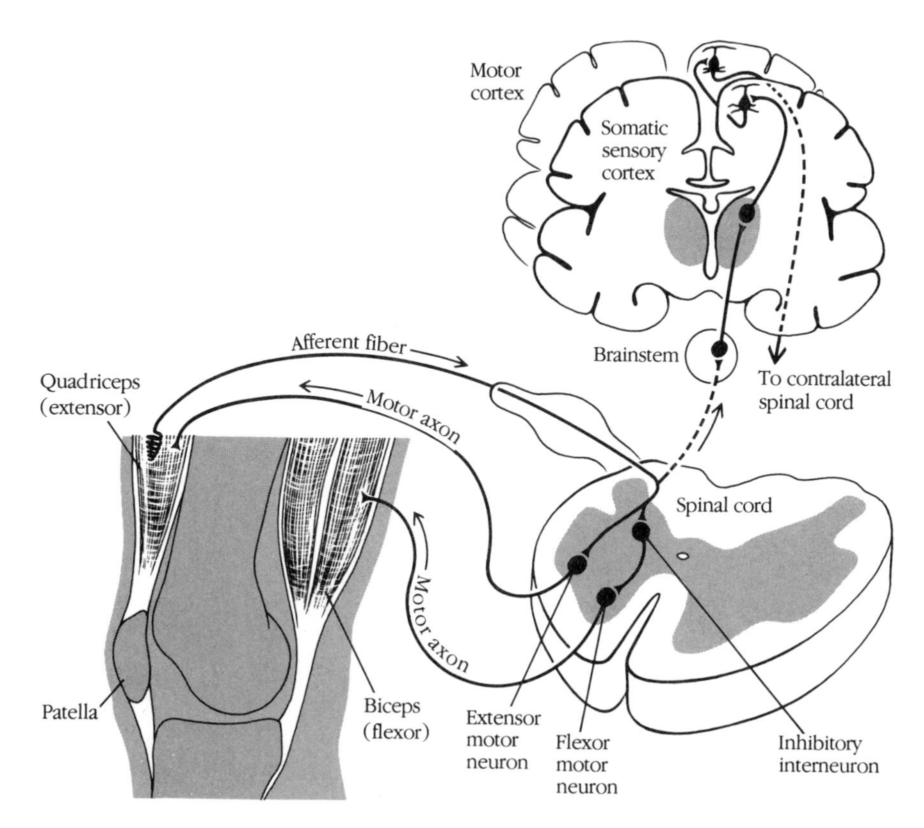


**FIGURE 5-3**
Human nervous system. (*a*) The central nervous system (black) comprises the brain and the spinal cord. Sensory and motor nerves (white) connect limbs and torso to the central nervous system. (*b*) The human brain. (*c*) Afferent (ascending) and efferent (descending) pathways between the knee and the brain via the spinal cord.

other organisms and with the environment are usually nonlinear phenomena, *rapid simulation is the only available approach to prediction and hence to behavior conducive to survival.*

## The Visual Cortex

The visual system is an ideal candidate for a nervous tissue simulator component, and its architecture is an ideal model for higher cortical functions generally. The capability of simulation by the visual system has been explored and explained by work on visual cortex of the monkey, cat, and frog (Hubel and Wiesel, 1974). The processing cortex contains as many as 20 layers of processing, each layer being a mapping of the next lower layer. Correlations in the nerve firing patterns in one layer determine the firing pattern of next layer. In this way, the *complex* cells achieve the ability to distinguish edges, motions, shapes, and so on. The ability to perceive velocity and even acceleration in visualized motion resides at a sufficiently high level in the layering, where so called *hypercomplex* cells individually recognize specific movements. This processing is rapid enough to serve as a rapid simulator (see Section 4-

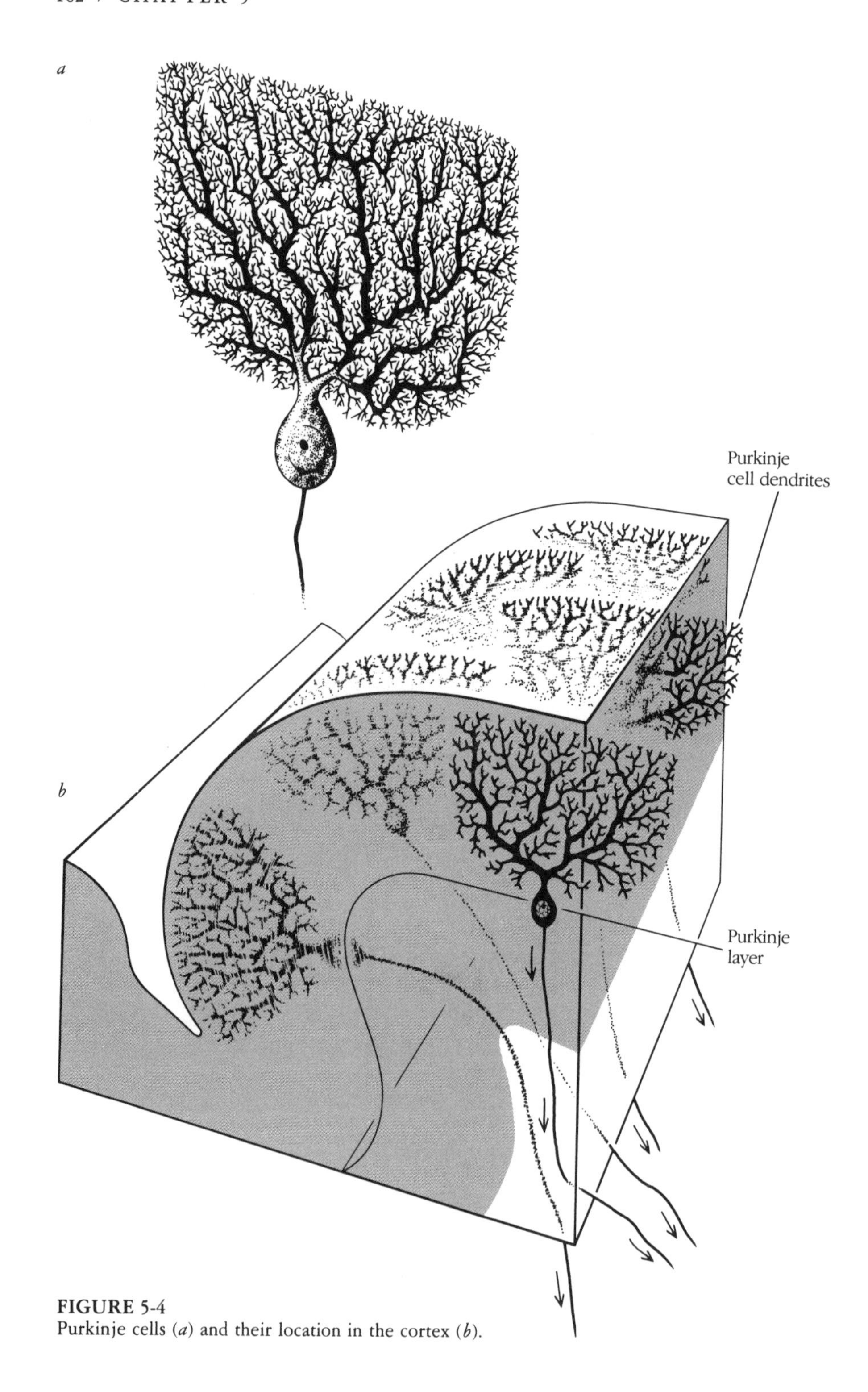


**FIGURE 5-4**
Purkinje cells (*a*) and their location in the cortex (*b*).

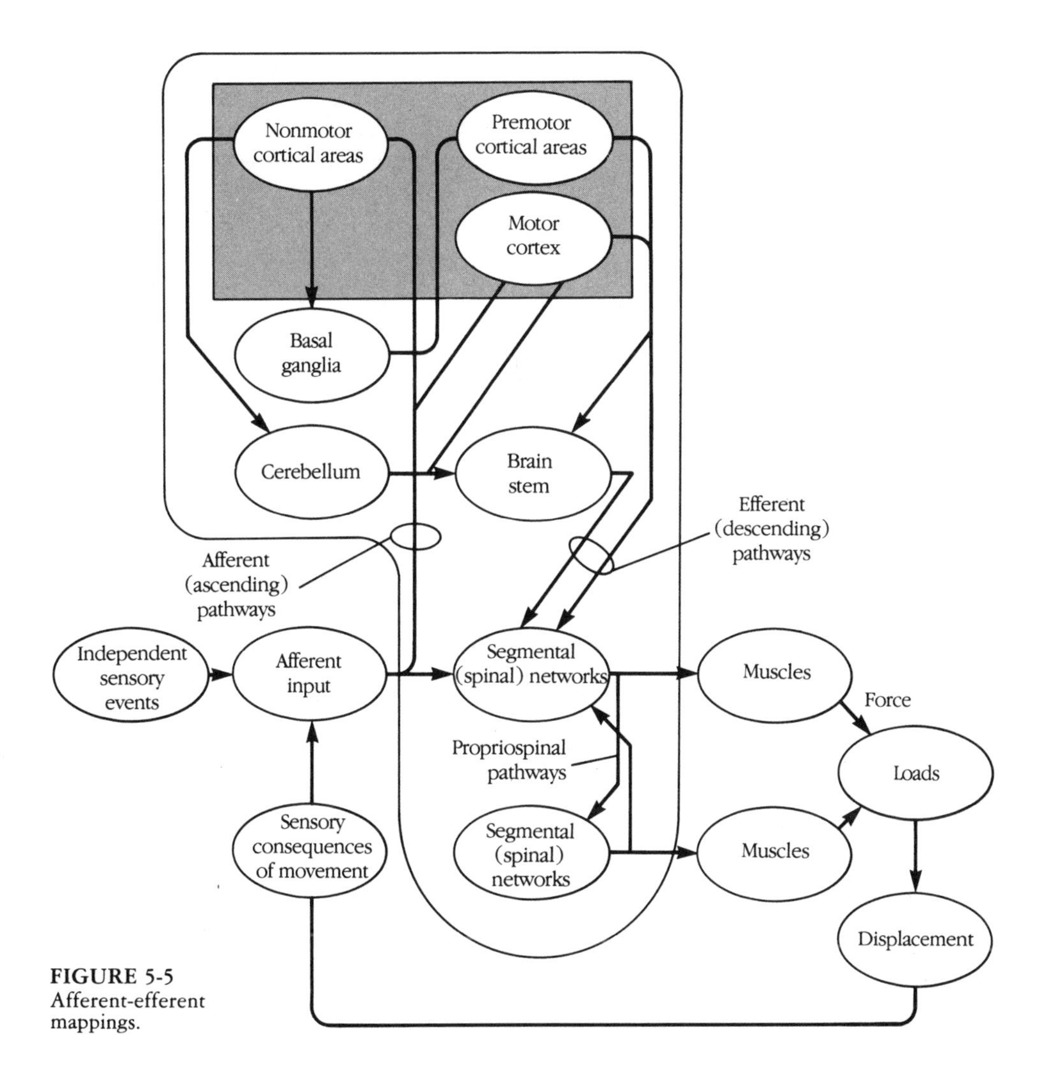


**FIGURE 5-5**
Afferent-efferent mappings.

10) if the visual system is integrated with the cerebral cortex. The understanding of how the layered structure of the visual cortex works has advanced greatly in recent years (Marr, 1982). In addition, researchers now know enough about neuronal connections to build model networks that exhibit simple behavior. One of the most beautiful examples of such models is Valentino Braitenberg's work on synthetic psychology (Braitenberg, 1984). These models show a variety of sensory-motor behavior using relatively simple components (Dewdney, 1987).

## Rapid Simulations

With the ability to perform rapid simulations, the nervous system can provide motor control with three distinct capabilities. The first, already present in the rudimentary sensory-motor reflexes of the Prechordata, is the so-called

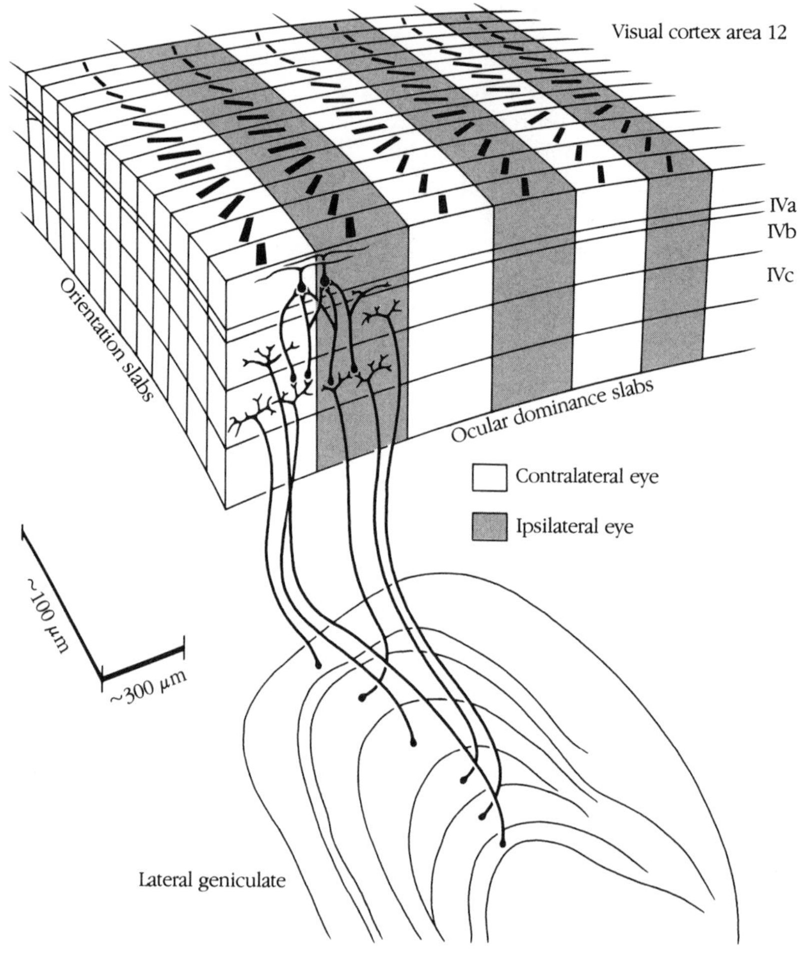


**FIGURE 5-6**
Visual cortex layers of a monkey.

simple behavior. In chordates, behavior need not be exclusively reflexive, although pure reflexive homeostatic mechanisms are retained by the chordate autonomic nervous system. The second capability is learning, in which pre-chordates have a rudimentary ability but chordates excel. Learning can take place over the entire life-span of an individual and clearly occurs without alteration of the genome, although gene activity may be involved. Finally, culture becomes a possibility for a sequence of generations of a species, which occurs especially in humans but also in many lower forms such as elephants.

## Brain Evolution

The evolution of the brain significantly altered the population dynamics of interacting species. The time required for cortical evolution has been very short compared with that required for the evolution of life, roughly 100 million years compared with 4 billion. The fastest rates of brain evolution have occurred in the last million years in higher primates. Brain tissue assembles during ontogeny from proteins that are coded and manufactured in a specific sequence by genetic control mechanisms. However, cultural factors as well as genetic ones may determine the survival of a species population. For example, different groups of humans with virtually identical genetic heritages have engaged in numerous battles for selection which are determined by culture, not by genetic mutation. Special weapons and organization skills have been determinant factors in these struggles for supremacy.

## Social Evolution

The mechanism of evolution is evolving as a natural consequence of the continuing evolution of energy coupling. Table 5-1 lists a number of steps in the evolution of energy flow through organic matter, their associated mechanisms for evolution, and their biochemical signatures.

The last entry in Table 5-1 suggests still another mechanism for evolution

**TABLE 5-1**
*Energy Flow Evolution*

| *Nature of energetics* | *Mechanism of evolution* | *Biochemical signature* |
|---|---|---|
| Coupling and transduction | Pregenetic proteinoid microspheres | Rudiments of glycolysis; pyrophosphate |
| Phosphate-driven polymerizations | Molecular genetics and Darwinian selection | ATP; electron transport chain |
| Regulation | Unicellular Darwinian selection | Cyclic AMP, $Ca^{2+}$, phosphokinases |
| Storage | Transition to mesozoans | Phosphagens: polyphosphates |
| Excitable tissues and rapid energy utilization | Metazoan Darwinian selection | Arginine-phosphate; muscle, bone, nervous tissue |
| Rapid simulation | Brain-based evolution determined by learning and culture | Creatine-phosphate; central nervous system |
| Secondary energy use to fuel social structures | Cultural behavior of populations; civilization | None; but many other signatures such as heat, electricity, coal, petroleum, fission |

beyond the nervous system. With brain, a new kind of life has emerged: multiindividual organisms. Humans are among them. However, at other stages in evolution organisms emerged that possess much less individual nervous tissue: for instance, the social insects, which use Arg-P as phosphagen. The diversity of collective behavior expressed by these species is astounding. In social insect communities, the cohesive force appears to be chemical. The simple nervous system is devoted to processing numerous chemical signals, along with touch and, usually, vision. In humans, the cohesive force seems most definitely to be embodied by the brain. Thoughts provide the bonding; they may be codified, replicated, translated, and mutated, as are the genes and gene products. In place of chromosomes with many genes, people have books that contain heritable information.

The emergent behavior of these metazoan collectives is predicated on the types of energy that the collective processes. For example, some ant colonies survive by cultivating fungi on green foliage harvested and shredded by worker ants. Bees can make their hive homeothermic, using their wings to evaporate water that they transport to the hive, whereas an individual bee is incapable of such a feat. In the final analysis, their lives revolve around building an energy reserve, honey.

The behavior of human collectives—such as clans, tribes, and nations—is also predicated on energy flux. Examples include fire for cooking, making pottery, and metal tools and weapons; sunlight for agriculature; windpower for seafaring commerce; motive force of steam engines for railroad transportation; combustive power of petroleum for locomotion such as aircraft flight; electricity for communication by telephone and television; and nuclear fission (and fusion) for power generation and war.

In the transitions from unicellular to multicellular organization and from prechordate to chordate organization, an energy-related signature is evident: the phosphagens Arg-P and Cr-P, respectively. It appears that a phosphagen refinement, from Arg-P to Cr-P, has profound consequences. The sequence of refinements in energy use by human collectives also appears to show an analogous importance. Consider, for example, the introduction of widespread electrical power. This power is different from the power of heat, sunlight, windpower, coal, and petroleum in that these other types of energy flux must be transduced into electricity to power telephones, television, automated printing, and electronic computers.

A refinement in cultural mechanisms has occurred with every refinement of energy flux coupling. Some steps have led to dramatic, emergent behavior. The practice of collective agriculture made an enormous impact, as did the advent of metallurgy and the bronze and iron ages. Recently, the nuclear age has followed the electricity age, closely and intimately. Suddenly, a new source of energy flux is available. Is it possible that the energy flux parameter has now increased to a point that could drive the nonlinear, dynamical process called civilization to chaos? Or is man's nervous system sufficiently advanced to predict future events and establish effective control mechanisms?

## REFERENCES

Braitenberg, V., *Vehicles, Experiments in Synthetic Psychology*, MIT Press, Cambridge, Mass., 1984.

Bullock, T. H., *Introduction to Nervous Systems*, W. H. Freeman, San Francisco, 1977.

Dewdney, A. K., "Computer Recreations," in *Scientific American* 256 (1987): March, 16–24.

Fox, T. O., "Evolution Evolving," in *Molecular Evolution and Protobiology*, edited by K. Matsuno, K. Dose, K. Harada, and D. L. Rohlfing, Plenum, New York, 1984.

Herrick, C. J., *Neurological Foundations of Animal Behavior*, Henry Holt, New York, 1922.

Hubel, D., and T. Wiesel, "Sequence, Regularity and Geometry of Orientation Columns in the Monkey Striate Cortex," *Journal of Comparative Neurology* 158 (1974): 267–294.

Kandel, E. R., and J. H. Schwartz, Eds., *Principles of Neural Science*, 2nd ed., Elsevier, New York, 1985.

Marr, D., *Vision*, W. H. Freeman, San Francisco, 1982.

Ricklefs, R. E., *Ecology*, Chiron Press, Newton, Mass., 1973.

# Glossary

**Definitions are restricted to those relevant to this book. Numbers in parentheses refer to sections in which a term appears in context in the book.**

*abiogenesis*: Synthesis of biological molecules by methods that do not include organisms but that are plausibly imputed to the primitive earth. (1-2)

*actin*: protein component of muscle. (3-3)

*activated monomer*: Monomer in an energy-rich state. In organisms, activation is achieved by chemical linkage to phosphate. (1-6, 2-1) *allostery*: regulation of protein function at one site caused by binding of effector molecule at another site such that a shape change takes place in the protein. (3-2)

*alpha particle* ($\alpha$): Nucleus of a helium atom, which contains two protons and two neutrons. (1-1)

*amino acid*: Monomeric constituent of proteins. Organisms use twenty different amino acids in gene directed protein synthesis. (1-3)

*amino acid adenylate*: Activated amino acid monomer. The activation involves a phosphate attached to ribose and adenine. (2-1)

*amino group*: part of an amino acid which arises from the simple molecule ammonia. (1-3)

*arthropod lineage*: One of two phylogenetic lineages used to classify multicellular organisms. It includes all the invertebrates. (The chordate lineage is the other.) (3-4)

*ATPase*: enzyme which catalyzes the hydrolysis or synthesis of ATP. (3-3)

*attractor*: Geometric structure of limited extent in phase space that is approached asymptotically by a trajectory representing the dynamics of a driven, dissipative nonlinear system. The attractor may be a fixed point, a limit cycle, or a complicated geometrical object with fractal structure called a *strange* attractor. If the motion on the attractor is extremely wild, the attractor is called a *chaotic* attractor. (4-5)

*autotroph*: Organism that has the metabolic sophistication needed to synthesize all its molecular constituents (including monomers, vitamins, and coenzymes) from very simple food-stuffs. (2-1, 2-3)

*axonal transport*: active energy utilizing process of molecule movement along the interior of an axon. (3-3)

*bacteriophage*: virus that multiplies in bacteria. (3-3)

*base*: When used as *the* bases in biology, the word refers to two purine bases (adenine and guanine) and three pyrimidine bases (cytosine, uracil, and thymine) that combine with a five-carbon sugar (ribose or deoxyribose) and a phosphate group to make nucleic acid, the monomer of DNA and RNA. (1-3)

*base pairing*: The formation of hydrogen bonds between two nucleic acids (bases) that have a specific chemical preference for each other: adenine with thymine (or with uracil in RNA) and cytosine with guanine. (2-1)

*beta decay*: process mediated by the weak fundamental force. The decay products include an electron and an antineutrino. In beta-plus decay the products are a positron (positively charged electron) and a neutrino. (1-1)

*bifurcation*: The splitting of a fixed-point attractor into a two-point attractor, or two into four, etc. Bifurcation is controlled by adjusting the parameters that control either the driving terms or the dissipative terms of a driven, dissipative dynamic system. (4-8)

*big bang model*: Currently dominant theory for the origin, from a point source, of the universe and all the matter-energy in it. (1-1)

*Boltzmann's formula*: Mathematical expression giving the relative probability for a fluctuation that changes the energy by the amount $E$ in a system in thermal equilibrium at temperature $T$: that is, $\exp(-E/k_B T)$, where $k_B$ is Boltzmann's constant. (1-6)

*calmodulin*: Calcium-binding regulatory protein found in many organisms. (3-3)

*carbon cycle*: Process in some stars that produces energy and synthesizes chemical elements. Also, in photosynthetic metabolism, the pathway responsible for the synthesis of carbohydrates (1-1) *carboxyl group*: part of an amino acid which arises from the simple molecule carbon dioxide. (1-3)

*catabolism*: metabolic degradation of foodstuffs. (3-2)

*cellular automaton*: Mathematical model that uses discrete time intervals to simulate the evolution of a system over time. Idea arose with Turing machines in the 1950s and has advanced with the development of computers. (4-10)

*chaos*: Disorder in a many-particle system. Also, behavior of a type of wild phase-space trajectory; it occurs when the associated dynamics possesses a positive Liapunov exponent. (2-4, 4-3, 4-5)

*chemiosmosis*: Mechanism by which the metabolic energy harvested by an electron transport chain is converted into ATP; it involves an electro-chemical state of the cellular membrane. (2-3)

*Chordata*: all the vertebrates and the proto-Chordata (sea-squirts, acorn worms, and lancelets). (3-4)

*chordate lineage*: One of two phylogenetic lineages used to classify multicellular organisms. It includes all the vertebrates and the Protochordata. (The arthropod lineage is the other.) (3-4)

*codon*: Triplet of three adjacent bases in DNA or in messenger RNA that code for an amino acid. A table of the genetic code lists the messenger RNA codons and their cognate amino acids. (2-2)

*coenzyme*: The part of many enzymes that possesses the catalytic activity of the enzyme; the remaining part is the *apoenzyme*; and together they constitute the *holoenzyme*. (1-4)

*conformational change*: shape change in a protein. (3-3)

*covalent bond*: strong, electron pair bond. (3-1)

*crosslinks*: intrachain connections in long polymers, such as disulfide bonds in proteins. (1-5)

*cytochrome*: Class of brightly colored proteins, containing heme iron, that are components of electron transport chains. (2-2)

*cytoskeleton*: structural components inside a cell that determine cell shape and cellular organelle movements. These components are made from either the proteins actin and myosin, as in muscle, or from the protein tubulin. (3-3)

*dehydration condensation*: Chemical linkage between monomers in biological polymers that removes one molecule of water. Such linkages can be broken by absorption of a molecule of water, a process called *hydrolysis*. (1-4)

DNA *(deoxyribonucleic acid)*: the molecular basis of genes. (1-3)

*driven dissipative system*: Any system with an energy-flow input and with a mechanism that degrades energy inputs into heat. (4-7, 5-2)

*effector*: protein binding molecule involved in regulating protein function. (3-2)

*electron transport chain*: Chain of reactions using cytochromes, iron-sulfur proteins, quinones, and other proteins to produce useful energy, mostly in the form of ATP in aerobic and photosynthetic metabolism. (2-2)

*emergent property*: Property of a structural level in a hierarchy that cannot be predicted from the properties of the components of the antecedent level (3-3)

*energy flow*: Changes in the dynamic state of matter caused by energy inputs. (1-1, 2-1, 2-4)

*energy transduction*: Conversion of one type of energy into another. Examples include conversion of oxidation-reduction energy into phosphate bond energy and conversion of light energy into oxidation-reduction energy. (2-3) *entropy*: Thermodynamic state function that measures the degree of disorder in a system at equilibrium. As heat is transferred to a system, its entropy increases. The second law of thermodynamics for an isolated system is stated in terms of the entropy. (2-4, 3-1)

*enzyme*: Biological catalyst. It is usually specific to a substrate and reaction and may or may not contain a coenzyme or inorganic compound that is the locus of catalytic activity. It may be regulated by binding special factors or by chemical modification catalyzed by other enzymes. (2-1)

*eukaryote*: (also eucaryote) Cell that has a membrane-enclosed nucleus, more than one chromosome, and organelles. Nearly all cells of the higher animals, fungi, protozoa, and many algae are eukaryotes, which probably evolved from prokaryotes. (3-4)

*evolution*: Literally an unfolding, meaning an opening out or a process of development, formation, or growth. This meaning is used in this book when discussing changes of a dynamic system over time. Otherwise, the more specialized biological meaning is intended: changes in species type over many generations. Darwin's theory of natural selection provided a mechanism for evolution which the study of molecular genetics later elucidated. This book expounds an underlying energy-related mechanism for evolution. (5-1)

*first messenger*: hormone or neurotransmitter substance which initiates activity in specific target cells (3-3)

*fixed point*: The simplest attractor (q.v.) for a driven, dissipative system. No matter how the system begins, its state asymptotically in time approaches the fixed point. Adjustment of parameters can cause the fixed point to bifurcate, become a limit cycle, or enlarge into a strange attractor. (4-4)

*free energy*: Thermodynamic state function that arises along with entropy and internal energy. There are two types: Helmholtz free energy for systems in contact with thermal reservoirs and Gibbs free energy for systems in contact with thermal and pressure reservoirs. *Free* means only that the energy is available for doing work. (3-1)

*free energy of formation*: Change in Gibbs free energy that attends the conversion of elements into a molecule. Molecules with highly negative free energies of formation are thermodynamically stable relative to the elements from which they are made. (1-2)

*furanose*: is the five-membered ring form of simple sugars. (1-4)

*genetic code*: Biochemical basis of heredity consisting of codons in DNA and RNA that determine the specific amino acid sequence in proteins. The code is essentially the same for the forms of life studied so far. (2-2, Table 2-2)

*glycolysis*: Glucose metabolism pathway that partially oxidizes glucose, converting it to pyruvate and transducing some of its free energy of formation into phosphate bond energy in ATP. It is perhaps the oldest form of energy transduction. (2-2)

*glycosidic linkage*: Dehydration condensation between two sugars (monosaccharides); it is hydrolyzed by water. (1-4)

*heterotroph*: An organism that lacks the metabolic sophistication necessary to synthesis all of its molecular constituents from very simple food stuffs and therefore needs to have some molecules in its diet. The required molecules are often amino acids and vitamins, which are components of essential coenzymes. Humans are heterotrophs. (2-1, 2-3) *histone*: A DNA-binding protein in eukaryotes. (3-3)

*hydrogen bond*: bond between two atoms created by sharing a proton. (3-1)

*hydrolysis*: The cleavage of a dehydration bond by water. All biological polymers can be degraded into their free monomeric constituents by hydrolysis. (1-6, 2-1)

*hydrophobic bond*: the association of nonpolar groups with each other in aqueous solution, arising because of the tendency of water molecules to exclude nonpolar molecules. (3-1)

*hydrosphere*: One of three components of the earth's surface environments (the others are the atmosphere and the lithosphere). The hydrosphere is made up of the oceans, lakes, rivers, clouds, and ice-caps. (1-2)

*intermediate neuron*: One of three major types of nerve cell. *Sensory* neurons provide input to the nervous system; *motor* neurons provide output from the nervous system to the muscle tissue; and *intermediate* neurons, which lie between, process the input into output. (5-4)

*iron catastrophe*: Event during a molten stage of earth (about 3.8 billion years ago) in which heavy elements such as iron melted and sank into the center of the earth, leaving behind the lighter siliceous crust and mantle. (1-2, 2-1, 2-3)

*kinase*: a phosphorylating enzyme. (3-2)

*Liapunov exponent*: The $\lambda$ in $\exp(\lambda t)$, . . . If two arbitrarily chosen trajectories begin from two initially very close together points in phase space, and if their early evolution results in exponentially diverging separation of the form: $\exp(\lambda t)$, then $\lambda$ is the Liapunov exponent. (4-3)

*limit cycle*: An attractor that is a closed curve. (4-4)

*lipid*: molecule containing long, nonpolar fatty acid chain and polar end, an important constituent of membranes. (3-1)

*lithosphere*: The earth's crust and the upper part of the mantle to which the crust is mechanically coupled. (1-2)

*macromolecule*: Any large polymer: protein, polynucleotides, or polysaccharide. (2-1)

*membrane*: Self-assembled two-dimensional surface containing lipids and proteins. The closed, outer envelope of a cell is a membrane, as is the eukaryotic nuclear envelope. Intact membrane structure is essential for chemiosmosis. (2-2)

*metastable*: Thermodynamically unstable molecule or complex which has a very slow natural rate of degradation. (3-1)

*microsphere*: Self-assembled spherical shell made of proteinoid, the boundary of which can display membranlike activity. Microspheres can serve as models for lifelike structures that may have evolved as precursors to primitive life. (2-3)

*microtubule*: Cytoskeleton component aggregated from tubulin protein molecules. (3-3)

*mitochondria*: ATP-generating organelles in eukaryotes. (3-1)

*mitosis*: Process during cellular reproduction by which the replicated chromosomes are segregated into daughter cells. (3-3)

*mixed oligomer*: Small polymers formed when monomers of different types, sometimes with phosphate, are polymerized (for example, trimers and tetramers). Coenzymes are important examples. (1-3)

*monomer*: Small molecule that can polymerize with others by dehydration linkage. The major types are amino acids, monosaccharides (sugars), vitamins, and nucleotides. (1-3)

*monosaccharide*: Simple sugar monomer. (1-3)
*morphogen*: Chemical substance that triggers morphological development in an embryo. Alan Turing invented morphogen models of gastrulation. (4-2)
*myosin*: Protein component of muscle. (3-3)

*neurotransmitter*: Molecule released by one nerve axion which triggers a response in a second nerve to which the first nerve is attached at a synapse. (3-3)
*nuclear fusion*: Primary energy-generating and element-generating process in stellar nucleosynthesis. (1-1)
*nucleoside*: A molecule similar to a nucleotide but lacking the phosphate group. (1-4, 2-1)
*nucleotide*: Monomeric precursor to polynucleotides (nuclear acids) such as DNA and RNA. It contains purine or pyrimidine bases, ribose, and phosphate. (1-4)

*oligosaccharide*: Small sugar polymer such as sucrose, a dimer. (1-4)
*oxidation-reduction reaction*: Chemical exchange of electrons, which may be accompanied by proton exchanges. (2-1)
*oxidative phosphorylation*: Production of a high-energy phosphate derivative of a substrate molecule using oxidation-reduction energy transduction to drive the reaction. (2-2)

*peptide*: Combination of two amino acids linking the amino group of one with the carboxyl group of the other. (1-3)
*peptide linkage*: Dehydration linkeage between two amino acids in a protein. (1-4, 2-1)
*period doubling*: Increase in the time it takes a trajectory to complete one cycle when a limit cycle divides into two interconnected cycles (a bifurcation).
*phase space*: Geometrical space defined by the $n$ variables that satisfy coupled, first-order in time, differential equations. It should not be confused with ordinary three-dimensional space. (4-3)
*phosphagen*: High-energy phosphate compounds used by organisms for energy storage to support the energy demands of excitable tissues. (3-4)
*phosphate*: Molecular form of phosphorus used by organisms. A molecule of phosphate contains four oxygen atoms and three hydrogen atoms covalently linked to phosphorus; it can possess four ionization states, depending on pH. At typical organismic pHs, phosphate is doubly charged, having lost two protons to ionization. (2-1)
*phosphate bond*: Linkage between a substrate molecule and a phosphate molecule; it may be energy-rich in some situations, such as in activated monomers. (2-1)
*phosphodiester bond*: Dehydration linkage between nucleotides in a polynucleotide. (1-5)
*phospholipid*: Lipid molecule with a polar end containing phosphate. (3-1)
*plate tectonics*: Study of geological processes based on the idea that large pieces (plates) of the earth's crust and upper mantle are slowly moving. Plates are created along an approximately 40,000 mile long oceanic rift containing thermal vents and are subducted along continent edges. Over millions of years, plates move hundreds of miles, collisions occur, and climate changes are dramatic. This idea has revolutionized geology. (1-2)
*polymers*: Large molecules made up of many monomers linked together by dehydration condensations. The major linear types are the proteins and the polynucleotides; the major branched type is the polysaccharides. (1-6)

*polynucleotides*: Macromolecular nucleic acids made of nucleotide monomers linked by phosphodiester bonds. The major types are DNA and RNA. (1-5, 2-1)

*power spectrum*: Result of a Fourier analysis of the time course of a phase-space trajectory for a dynamic system. In linear systems, the power spectrum provides the values of the fundamental frequencies. In nonlinear systems, the power spectrum also reveals overtones and harmonic mixtures. (4-5)

*Prechordata*: All multicellular invertebrates. (3-4)

*predictability*: Quality or state of being predictable. In this book, the problem of forecasting the outcome of nonlinear dynamic events. Its solution requires rapid simulation. (4-8, 4-10)

*primordial dozen*: The set of twelve chemical elements from which a hypothetical model for minimal cellular life can be constructed. (1-1)

*prokaryote (also procaryote)*: Small, single-celled organism having only a single membrane, the cell envelope. They include bacteria, blue-green algae, spirochetes, rickettsiae, and pleuropneumonialike organisms. They possess only one chromosome. (3-4)

*proteins*: Macromolecules containing many amino acid monomers connected by peptide bonds. (1-5, 2-1)

*proteinoid*: An abiogenically produced polymer of amino acids made by dry heating of amino acid mixtures. It serves as model compound for primitive proteins. (2-1, 2-3)

*Protochordata*: Three types of organism believed to be the precursors to chordates: sea-squirts, acorn worms, and lancelets. (3-4)

*Purkinje cell*: Prominent cerebellar intermediary neuron. (5-4)

*pyrophosphate*: Phosphate dimer molecule, the simplest example of a phosphodiester bond and of a high-energy phosphate bond. It may be the form of phosphate energy used by primitive life before ATP evolved. (1-4)

*rapid simulator*: In nonlinear dynamics, a mechanism that simulates the future on a faster time scale than that which exists for the real time system being simulated. (4-10)

*red giant*: is a large, cool star of high luminosity. (1-1)

*regulation*: Control of enzyme binding of substrates and their catalytic action. (3-2)

*residues*: Distinctive chemical groups in amino acids. (3-1)

*resonant*: reactions or interactions in which the frequencies of two different reactants are equal. (1-1)

*RNA (ribonucleic acid)*: Molecular basis of gene transcription and translation. mRNA is *m*essenger RNA, the transcription product of genes. tRNA is *t*ransfer RNA used in translation of mRNA into protein, a process coordinated by ribosomes, which also contain rRNA, i.e., *r*ibosomal RNA. (1-3)

*second law of thermodynamics*: The statement that for an isolated system at constant energy, the equilibrium state is the state of maximum entropy. For systems in contact with a thermal reservoir, the second law states that the Helmholtz free energy is a minimum at equilibrium; and for systems in contact with both a thermal reservoir and a pressure reservoir, the Gibbs free energy is minimum at equilibrium. (2-4, 3-1)

*second messenger*: Molecule (cyclic AMP or cyclic GMP), the production of which is triggered by first messenger, that triggers cellular response. (3-3) *self-assembly*: Process by which molecules recognize each other and aggregate into complexes and even into membranes. The process exhibits specificity and is driven by entropy changes in the associated water molecules. (3-1)

*stellar nucleosynthesis*: Mechanism of formation of the chemical elements in stars. (1-1)

*strong nuclear force*: One of the four fundamental forces in nature (the others being the gravitational, the weak, and the electromagnetic). Strong nuclear forces are over 2000 times stronger than electromagnetism, the second strongest of the forces. (1-1)

*subunit*: In an enzyme, a chain in a complex of several polypeptide chains. (3-1)

*synthetase*: Enzyme responsible for the specific attachment of an amino acid to its cognate tRNA. (2-3)

*tetramer*: Protein complex made from four subunits. (3-2)

*transcription*: Process of converting the sequence of bases of a gene into the complementary sequence in a messenger RNA (mRNA); it is the first step in gene-directed protein biosynthesis. (2-2)

*translation*: Process of converting the base sequence of mRNA into the amino acid sequence of a protein. (2-2)

*translational motion*: The motion caused by linear kinetic energy. (3-1)

*tricarboxylic acid cycle*: (Also citric acid cycle and Krebs cycle): Pathway in energy metabolism that extracts electrons from pyruvate and feeds them to the electron transport chain. (2-2)

*uroboros*: Mythical serpent symbolizing the idea of self-begetting. (Introduction, 1-6, 2-1, 2-3, 2-4)

*urRNA*: Short RNA molecules used in this book as the model for primitive molecular genetics at the beginning of life. (2-3)

*van der Waals bond*: Weak short-range force between two molecules caused by induced dipoles. (3-1)

*variational principle*: Physical principle in which some quantity maintains an external (maximum or minimum) value. (3-3)

*ylem*: Primordial material at the start of the expansion of the universe which we now observe billions of years after the original "big bang." (1-1)

# Index